8.50
i/p 2/07

HEINEMANN MODULAR MATHEMATICS
for
EDEXCEL AS AND A-LEVEL
Pure Mathematics 3

Geoff Mannall   Michael Kenwood

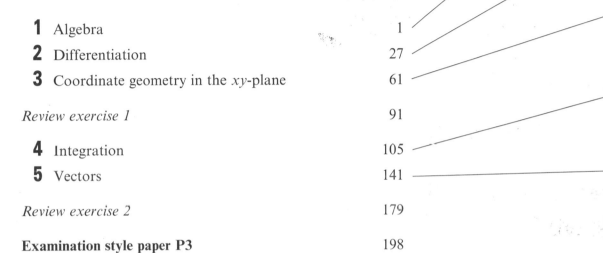

Heinemann

Edexcel
Success through qualifications

Heinemann Educational Publishers,
a division of Heinemann Publishers (Oxford) Ltd,
Halley Court, Jordan Hill, Oxford, OX2 8EJ

OXFORD MELBOURNE AUCKLAND JOHANNESBURG
BLANTYRE IBADAN GABORONE PORTSMOUTH NH (USA)
CHICAGO

First published 2000

02 01 10 9 8 7 6 5 4 3

ISBN 0 435 51090 8

Cover design by Gecko Limited

Original design by Geoffrey Wadsley; additional design work by Jim Turner

Typeset and illustrated by Tech-Set Limited, Gateshead, Tyne and Wear

Printed in Great Britain by The Bath Press, Bath

Acknowledgements:

The publisher's and authors' thanks are due to Edexcel for permission to
reproduce questions from past examination papers. These are marked with an [E].
   The answers have been provided by the authors and are not the responsibility
of the examining board.

About this book

This book is designed to provide you with the best preparation possible for your Edexcel P3 exam. The series authors are senior examiners and exam moderators themselves and have a good understanding of Edexcel's requirements.

Use this **new edition** to prepare for the new 6-unit specification. Use the first edition (*Heinemann Modular Mathematics for London AS and A-level*) if you are preparing for the 4-module syllabus.

Finding your way around

To help to find your way around when you are studying and revising use the:

- **edge marks** (shown on the front page) – these help you to get to the right chapter quickly;
- **contents list** – this lists the headings that identify key syllabus ideas covered in the book so you can turn straight to them;
- **index** – if you need to find a topic the **bold** number shows where to find the main entry on a topic.

Remembering key ideas

We have provided clear explanations of the key ideas and techniques you need throughout the book. Key ideas you need to remember are listed in a **summary of key points** at the end of each chapter and marked like this in the chapters:

$$\blacksquare \qquad \frac{d}{dx}(\sin x) = \cos x$$

Exercises and exam questions

In this book questions are carefully graded so they increase in difficulty and gradually bring you up to exam standard.

- **past exam questions** are marked with an [E];
- **review exercises** on pages 91 and 179 help you practise answering questions from several areas of mathematics at once, as in the real exam;
- **exam style practice paper** – this is designed to help you prepare for the exam itself;
- **answers** are included at the end of the book – use them to check your work.

Contents

Algebra

In chapter 1 of Book P2 you learned how to process algebraic rational expressions by using addition, subtraction, multiplication and division.

For example, you learned how to add $\dfrac{2}{x+5} + \dfrac{3}{x+4}$ by using the common denominator $(x+5)(x+4)$ to give

$$\frac{2(x+4) + 3(x+5)}{(x+5)(x+4)}$$

$$= \frac{2x+8+3x+15}{(x+5)(x+4)} = \frac{5x+23}{(x+5)(x+4)}$$

Now you need to be able to *start* with an expression like

$$\frac{5x+23}{(x+5)(x+4)}$$

and split it into

$$\frac{2}{x+5} + \frac{3}{x+4}$$

This process of taking a single fraction and breaking it up into the sum (or difference) of two or more fractions is known as **splitting an expression into partial fractions**. It has many applications, including differentiation, integration and in expansions using the binomial series, as you will learn later in this book.

1.1 Partial fractions

Before you can decompose an expression into partial fractions, you must be able to recognise the different types which can occur. In all cases, it is essential to factorise the denominator successfully. Here you will see how to deal with each type by considering specific examples.

Linear factors in the denominator

When you need to split an expression into its partial fractions, the first thing you must do is to factorise the denominator of the given expression. With, for example, the fraction $\dfrac{5x+1}{x^2+x-2}$ you first rewrite it as

$$\frac{5x+1}{(x-1)(x+2)}$$

■ **For a fraction with only linear factors in the denominator, and where the degree of the denominator exceeds that of the numerator, e.g. $\dfrac{5x+1}{(x-1)(x+2)}$, the partial fractions are of the form**

$$\frac{A}{x-1}+\frac{B}{x+2}$$

where A and B are constants.

So:
$$\frac{5x+1}{(x-1)(x+2)} \equiv \frac{A}{x-1}+\frac{B}{x+2}$$

Note that this is an identity, because when you have found the values of A and B, $\dfrac{5x+1}{(x-1)(x+2)}$ will be equal to $\dfrac{A}{x-1}+\dfrac{B}{x+2}$ for *all* values of x, and not just for *some* values of x.

Now:
$$\frac{A}{x-1}+\frac{B}{x+2} \equiv \frac{A(x+2)+B(x-1)}{(x-1)(x+2)}$$

So:
$$\frac{5x+1}{(x-1)(x+2)} \equiv \frac{A(x+2)+B(x-1)}{(x-1)(x+2)}$$

However, if these two fractions are identical, and also the denominators of these two fractions are identical, then the numerators of the two fractions must also be identical. That is:

$$5x+1 \equiv A(x+2)+B(x-1)$$

The values of A and B can now be found by selecting convenient values of x which make one of the terms on the right-hand side equal to zero, like this:

$x=1 \Rightarrow$
$$6 = 3A$$
$$A = 2$$

$x=-2 \Rightarrow$
$$-9 = -3B$$
$$B = 3$$

So:
$$\frac{5x+1}{(x-1)(x+2)} \equiv \frac{2}{x-1}+\frac{3}{x+2}$$

and this can easily be checked by adding $\dfrac{2}{x-1}$ and $\dfrac{3}{x+2}$

Example 1

Express $\dfrac{4x-2}{x^3-x}$ in partial fractions.

$$\frac{4x-2}{x^3-x} = \frac{4x-2}{x(x^2-1)} \qquad \text{(take out the common factor } x\text{)}$$

$$= \frac{4x-2}{x(x-1)(x+1)} \qquad \text{(factorise the difference of two squares)}$$

Since each of the three factors in the denominator is linear:

$$\frac{4x-2}{x(x-1)(x+1)} \equiv \frac{A}{x} + \frac{B}{x-1} + \frac{C}{x+1}$$

Now: $\quad \dfrac{A}{x} + \dfrac{B}{x-1} + \dfrac{C}{x+1} \equiv \dfrac{A(x-1)(x+1) + Bx(x+1) + Cx(x-1)}{x(x-1)(x+1)}$

So: $\qquad 4x-2 \equiv A(x-1)(x+1) + Bx(x+1) + Cx(x-1)$

Choosing values of x which make two of the terms on the right-hand side zero:

$x = 0 \Rightarrow \qquad\qquad\qquad -2 = -A$

$\qquad\qquad\qquad\qquad\qquad A = 2$

$x = 1 \Rightarrow \qquad\qquad\qquad 2 = 2B$

$\qquad\qquad\qquad\qquad\qquad B = 1$

$x = -1 \Rightarrow \qquad\qquad\quad -6 = 2C$

$\qquad\qquad\qquad\qquad\qquad C = -3$

Thus: $\qquad\qquad \dfrac{4x-2}{x^3-x} \equiv \dfrac{2}{x} + \dfrac{1}{x-1} - \dfrac{3}{x+1}$

Example 2

Express $\dfrac{3x^2+7x-8}{(x+1)(x+2)(x-3)}$ in partial fractions.

Since the denominator has factors that are each linear, and the degree of the numerator is 2 which is less than the degree of the denominator, the form of the partial fractions is:

$$\frac{A}{x+1} + \frac{B}{x+2} + \frac{C}{x-3} \equiv \frac{A(x+2)(x-3) + B(x+1)(x-3) + C(x+1)(x+2)}{(x+1)(x+2)(x-3)}$$

Therefore, equating numerators:

$$3x^2 + 7x - 8 \equiv A(x+2)(x-3) + B(x+1)(x-3) + C(x+1)(x+2)$$

$$x = -1 \Rightarrow \qquad\qquad 3 - 7 - 8 = -4A$$
$$-12 = -4A$$
$$A = 3$$

$$x = -2 \Rightarrow \qquad\qquad 12 - 14 - 8 = 5B$$
$$-10 = 5B$$
$$B = -2$$

$$x = 3 \Rightarrow \qquad\qquad 27 + 21 - 8 = 20C$$
$$40 = 20C$$
$$C = 2$$

Therefore: $\qquad \dfrac{3x^2 + 7x - 8}{(x+1)(x+2)(x-3)} \equiv \dfrac{3}{x+1} - \dfrac{2}{x+2} + \dfrac{2}{x-3}$

Quadratic factor in the denominator

If any of the factors in the denominator is *not* linear then the partial fractions cannot take the form shown in the previous section.

■ **For a fraction that has a non-reducible quadratic factor in the denominator, and where the degree of the denominator exceeds that of the numerator, such as**

$$\frac{x^2 - 5x + 1}{(x^2 + 1)(x - 2)}$$

the partial fractions are of the form

$$\frac{Ax + B}{x^2 + 1} + \frac{C}{x - 2}$$

where A, B and C are constants.

It is *essential* to ensure that each partial fraction with a quadratic denominator has a numerator of the form $Ax + B$. Any numerator of the form A will not, in general, work if the denominator is quadratic.

Example 3

Express $\dfrac{5x^2 + 4x + 4}{(x+2)(x^2+4)}$ in partial fractions.

Since one of the factors in the denominator is quadratic the partial fractions are of the form

$$\frac{A}{x+2} + \frac{Bx + C}{x^2 + 4} \equiv \frac{A(x^2 + 4) + (x+2)(Bx + C)}{(x+2)(x^2+4)}$$

So, equating numerators:

$$5x^2 + 4x + 4 \equiv A(x^2 + 4) + (x + 2)(Bx + C)$$

$$x = -2 \Rightarrow \qquad 20 - 8 + 4 = 8A$$
$$16 = 8A$$
$$A = 2$$

Since there is no real value of x that will make the term $A(x^2 + 4)$ equal to zero, we now equate coefficients.

Equating coefficients of x^2:
$$5 = A + B$$
$$5 = 2 + B$$
$$B = 3$$

Equating coefficients of x:
$$4 = 2B + C$$
$$4 = 6 + C$$
$$C = -2$$

Thus:
$$\frac{5x^2 + 4x + 4}{(x + 2)(x^2 + 4)} \equiv \frac{2}{x + 2} + \frac{3x - 2}{x^2 + 4}$$

You must make sure that you have factorised the denominator of the given fraction *completely*, before you try to express it in partial fractions. If the denominator is $(x^2 - 3x + 2)(x^2 - x + 3)$ then the partial fractions are *not* of the form

$$\frac{Ax + B}{x^2 - 3x + 2} + \frac{Cx + D}{x^2 - x + 3}$$

because $(x^2 - 3x + 2)(x^2 - x + 3)$ is not completely factorised. It can be factorised further into $(x - 1)(x - 2)(x^2 - x + 3)$. So the partial fractions are of the form

$$\frac{A}{x - 1} + \frac{B}{x - 2} + \frac{Cx + D}{x^2 - x + 3}$$

Example 4

Express

$$\frac{-2x - 1}{(x^2 - 3x + 2)(x^2 - x + 3)}$$

in partial fractions.

Since the denominator can be factorised to $(x - 1)(x - 2)(x^2 - x + 3)$ the partial fractions are of the form

$$\frac{A}{x - 1} + \frac{B}{x - 2} + \frac{Cx + D}{x^2 - x + 3} \equiv \frac{A(x - 2)(x^2 - x + 3) + B(x - 1)(x^2 - x + 3) + (Cx + D)(x - 1)(x - 2)}{(x - 1)(x - 2)(x^2 - x + 3)}$$

Equating numerators:

$$-2x - 1 \equiv A(x-2)(x^2 - x + 3) + B(x-1)(x^2 - x + 3) + (Cx + D)(x-1)(x-2)$$

$x = 1 \Rightarrow$
$$-3 = A(-1)(1 - 1 + 3)$$
$$-3 = -3A$$
$$A = 1$$

$x = 2 \Rightarrow$
$$-5 = B(1)(4 - 2 + 3)$$
$$-5 = 5B$$
$$B = -1$$

Equating coefficients of x^3:
$$0 = A + B + C$$
$$0 = 1 - 1 + C$$
$$0 = C$$

Equating constant terms:
$$-1 = -6A - 3B + 2D$$
$$-1 = -6 + 3 + 2D$$
$$2D = 2$$
$$D = 1$$

So:
$$\frac{-2x - 1}{(x^2 - 3x + 2)(x^2 - x + 3)} \equiv \frac{1}{x - 1} - \frac{1}{x - 2} + \frac{1}{x^2 - x + 3}$$

Repeated factor in the denominator

If the factors in the denominator include one that is repeated, e.g. $(x + 4)^2$, then the partial fractions take yet another form, which you must remember.

■ **For a fraction that has a repeated factor in the denominator, and where the degree of the denominator exceeds that of the numerator, e.g.**

$$\frac{2(x^2 - 2x - 1)}{(x + 1)(x - 1)^2}$$

the partial fractions are of the form

$$\frac{A}{x + 1} + \frac{B}{x - 1} + \frac{C}{(x - 1)^2}$$

where A, B and C are constants.

Example 5

Express $\dfrac{2(x^2 - 2x - 1)}{(x + 1)(x - 1)^2}$ in partial fractions.

The partial fractions are of the form

$$\frac{A}{x + 1} + \frac{B}{x - 1} + \frac{C}{(x - 1)^2} \equiv \frac{A(x - 1)^2 + B(x + 1)(x - 1) + C(x + 1)}{(x + 1)(x - 1)^2}$$

Equating numerators gives

$$2(x^2 - 2x - 1) \equiv A(x-1)^2 + B(x+1)(x-1) + C(x+1)$$

$x = -1 \Rightarrow \qquad\qquad 2(1 + 2 - 1) = 4A$

$$4 = 4A$$
$$A = 1$$

$x = 1 \Rightarrow \qquad\qquad 2(1 - 2 - 1) = 2C$

$$-4 = 2C$$
$$C = -2$$

Equating constant terms: $\qquad\qquad -2 = A - B + C$

$$-2 = 1 - B - 2$$
$$B = 1$$

So: $\qquad \dfrac{2(x^2 - 2x - 1)}{(x+1)(x-1)^2} \equiv \dfrac{1}{x+1} + \dfrac{1}{x-1} - \dfrac{2}{(x-1)^2}$

Example 6

Express $\dfrac{x^2 + 3x + 4}{(x+2)^3}$ in partial fractions.

The partial fractions are of the form:

$$\frac{A}{x+2} + \frac{B}{(x+2)^2} + \frac{C}{(x+2)^3}$$

$$\equiv \frac{A(x+2)^2 + B(x+2) + C}{(x+2)^3}$$

Equating numerators gives

$$x^2 + 3x + 4 \equiv A(x+2)^2 + B(x+2) + C$$

$x = -2 \Rightarrow \qquad\qquad 4 - 6 + 4 = C$

$$C = 2$$

Equating coefficients of x^2: $\qquad\qquad 1 = A$

Equating coefficients of x: $\qquad\qquad 3 = 4A + B$

$$3 = 4 + B$$
$$B = -1$$

So: $\qquad \dfrac{x^2 + 3x + 4}{(x+2)^3} \equiv \dfrac{1}{x+2} - \dfrac{1}{(x+2)^2} + \dfrac{2}{(x+2)^3}$

Improper algebraic fractions

The polynomial $3x + 2$ is of degree 1 and the polynomial $x^2 - 5x + 2$ is of degree 2. The degree of a polynomial in x is the same as the degree of the highest power of x in the polynomial.

A fraction where the degree of the numerator is less than the degree of the denominator is called a **proper fraction**. A fraction where the degree of the numerator is either equal to the degree of the denominator or higher than the degree of the denominator is called an **improper fraction**.

So

$$\frac{2x + 3}{3x^2 - 2x + 1}$$

is a proper fraction because the numerator has degree 1 and the denominator has degree 2. But

$$\frac{3x^2 + 2x + 6}{5x^2 - 7}$$

is an improper fraction because the numerator has degree 2, which is equal to the degree of the denominator, for this is also of degree 2. Similarly,

$$\frac{2x^3 + 3x^2 - 7}{5x - 3}$$

is an improper fraction because the numerator is of degree 3, which is greater than the degree of the denominator, for this is of degree 1.

When you try to split a fraction into its component partial fractions you must make sure that it is a proper fraction. If it is improper, you must first divide the denominator into the numerator before you try to split up the fraction into its partial fractions.

Example 7

Express $\dfrac{2x^2 + 8x + 7}{(x + 2)(x + 3)}$ in partial fractions.

Since the degree of both the numerator and the denominator is 2, the fraction is improper. So the denominator must be divided into the numerator.

First multiply out the denominator:

$$\frac{2x^2 + 8x + 7}{(x + 2)(x + 3)} \equiv \frac{2x^2 + 8x + 7}{x^2 + 5x + 6}$$

$$\begin{array}{r} 2 \\ x^2 + 5x + 6 \overline{)2x^2 + 8x + 7} \\ \underline{2x^2 + 10x + 12} \\ -2x - 5 \end{array}$$

So $x^2 + 5x + 6$ divides into $2x^2 + 8x + 7$ twice and leaves a remainder $-2x - 5$.

Now when 33 is divided by 7, the quotient is 4 with a remainder 5. This answer can be written $4\frac{5}{7}$, that is $4 + \frac{5}{7}$. In the same way,

$$\frac{2x^2 + 8x + 7}{x^2 + 5x + 6}$$

can be written:

$$2 + \frac{-2x - 5}{x^2 + 5x + 6} \equiv 2 + \frac{-(2x + 5)}{x^2 + 5x + 6}$$

$$\equiv 2 - \frac{2x + 5}{x^2 + 5x + 6}$$

The fraction

$$\frac{2x + 5}{x^2 + 5x + 6}$$

is proper since the degree of the numerator (1) is less than the degree of the denominator (2). It can therefore now be split into partial fractions.

$$\frac{2x + 5}{x^2 + 5x + 6} \equiv \frac{2x + 5}{(x + 2)(x + 3)}$$

The partial fractions are of the form

$$\frac{A}{x + 2} + \frac{B}{x + 3} \equiv \frac{A(x + 3) + B(x + 2)}{(x + 2)(x + 3)}$$

Equating numerators gives:

$$2x + 5 \equiv A(x + 3) + B(x + 2)$$

$x = -2 \Rightarrow \qquad\qquad 1 = A$

$x = -3 \Rightarrow \qquad\qquad -1 = -B$

$$B = 1$$

So: $\qquad \dfrac{2x^2 + 8x + 7}{(x + 2)(x + 3)} \equiv 2 - \dfrac{2x + 5}{x^2 + 5x + 6}$

$$\equiv 2 - \left[\frac{1}{x + 2} + \frac{1}{x + 3}\right]$$

$$\equiv 2 - \frac{1}{x + 2} - \frac{1}{x + 3}$$

Example 8

Express $\dfrac{x^3 + 3x^2 + 10}{(x+1)(x+4)}$ in partial fractions.

The degree of the numerator (3) is higher than the degree of the denominator (2), so the fraction is improper.

$$\frac{x^3 + 3x^2 + 10}{(x+1)(x+4)} \equiv \frac{x^3 + 3x^2 + 10}{x^2 + 5x + 4}$$

So by division:

$$
\begin{array}{r}
x - 2 \\
x^2 + 5x + 4 \,\overline{)\,x^3 + 3x^2 +\ 0x + 10} \\
\underline{x^3 + 5x^2 +\ 4x} \\
-2x^2 -\ 4x + 10 \\
\underline{-2x^2 - 10x -\ 8} \\
6x + 18
\end{array}
$$

That is, $\quad x^3 + 3x^2 + 10 \equiv (x^2 + 5x + 4)(x - 2) + 6x + 18$

So: $\qquad \dfrac{x^3 + 3x^2 + 10}{x^2 + 5x + 4} \equiv x - 2 + \dfrac{6x + 18}{x^2 + 5x + 4}$

$$\frac{x^3 + 3x^2 + 10}{(x+1)(x+4)} \equiv x - 2 + \frac{6x + 18}{(x+1)(x+4)}$$

Now $\dfrac{6x + 18}{(x+1)(x+4)}$ is proper, since the degree of the

numerator (1) is less than the degree of the denominator (2). We can therefore split it into partial fractions.

$$\frac{6x + 18}{(x+1)(x+4)} \equiv \frac{A}{x+1} + \frac{B}{x+4}$$

$$\equiv \frac{A(x+4) + B(x+1)}{(x+1)(x+4)}$$

Equating numerators gives

$$6x + 18 \equiv A(x+4) + B(x+1)$$

$x = -1 \Rightarrow$
$$12 = 3A$$
$$A = 4$$

$x = -4 \Rightarrow$
$$-6 = -3B$$
$$B = 2$$

So: $\qquad \dfrac{x^3 + 3x^2 + 10}{(x+1)(x+4)} \equiv x - 2 + \dfrac{4}{x+1} + \dfrac{2}{x+4}.$

Exercise 1A

Express as partial fractions:

1 $\dfrac{2x+5}{(x+2)(x+3)}$ **2** $\dfrac{2x+2}{(x-1)(x+3)}$

3 $\dfrac{x+1}{(x+3)(x+4)}$ **4** $\dfrac{x+7}{x^2+5x+6}$

5 $\dfrac{2x^2+12x-10}{(x-1)(2x-1)(x+3)}$ **6** $\dfrac{3x^2-x+6}{(x^2+4)(x-2)}$

7 $\dfrac{x^2-2x+9}{(x^2+3)(x-3)}$ **8** $\dfrac{-2x^2+4x-4}{(x^2+5)(2x+3)}$

9 $\dfrac{-6x^2+x-12}{(5+2x^2)(x+3)}$ **10** $\dfrac{-2x^2+13}{(2x+1)(x^2+2x+7)}$

11 $\dfrac{2x-7}{(x-5)^2}$ **12** $\dfrac{x^2+4x+7}{(x+3)^3}$

13 $\dfrac{-3x^2+10x+5}{(x+2)(x-1)^2}$ **14** $\dfrac{-5x^2+8x+9}{(x+2)(x-1)^2}$

15 $\dfrac{10x+9}{(2x+1)(2x+3)^2}$ **16** $\dfrac{x}{x-1}$

17 $\dfrac{x^2}{x-1}$ **18** $\dfrac{x^2+1}{x^2-1}$

19 $\dfrac{x^2+2}{x(x-1)}$ **20** $\dfrac{x^3}{x^2-1}$

21 $\dfrac{9-2x-2x^2}{(1+x)(2-x)}$ **22** $\dfrac{4x^2-3x+2}{2x^2-x-1}$

23 $\dfrac{2x^3+10x^2+12x+1}{(x+2)(x+3)}$ **24** $\dfrac{x^3+x^2-2x+4}{x^2-4}$

25 $\dfrac{-x^4-x^3+2x^2-x-2}{x^2(x+1)}$ **26** $\dfrac{13}{(2x-3)(3x+2)}$

27 $\dfrac{4x^2+5x+9}{(2x-1)(x+2)^2}$ **28** $\dfrac{x^3+4x^2+3x+4}{(x^2+1)(x+1)^2}$

29 $\dfrac{4x^4+6x^3+4x^2+x-3}{x^2(2x+3)}$ **30** $\dfrac{4x+3}{(2x-1)(3x+1)}$

31 $\dfrac{x^3+3x^2-2x-5}{(x-1)^2(x^2+2)}$ **32** $\dfrac{3x^2+12x+8}{(2x+3)(x^2-4)}$

33 $\dfrac{x^3-x^2-1}{x(x^2+x+1)}$ **34** $\dfrac{x^2+2x+3}{x^2(x+1)}$ **35** $\dfrac{3}{x^3+1}$

1.2 The remainder theorem and the factor theorem

When a polynomial is divided by a linear expression there is usually a remainder. For example,

$$(3x^4 - 7x^3 - 10x^2 + 19x - 19) \div (x - 3)$$

$$
\begin{array}{r}
3x^3 + 2x^2 - 4x + 7 \\
x - 3 \overline{)3x^4 - 7x^3 - 10x^2 + 19x - 19} \\
\underline{3x^4 - 9x^3} \\
2x^3 - 10x^2 \\
\underline{2x^3 - 6x^2} \\
-4x^2 + 19x \\
\underline{-4x^2 + 12x} \\
7x - 19 \\
\underline{7x - 21} \\
2
\end{array}
$$

So: $\qquad (3x^4 - 7x^3 - 10x^2 + 19x - 19) \div (x - 3)$

gives a quotient of $3x^3 + 2x^2 - 4x + 7$ and a remainder of 2. You can write this sentence mathematically as:

$$\frac{3x^4 - 7x^3 - 10x^2 + 19x - 19}{x - 3} \equiv 3x^3 + 2x^2 - 4x + 7 + \frac{2}{x - 3}$$

Now if both sides of the identity are multiplied by $(x - 3)$ then

$$\frac{3x^4 - 7x^3 - 10x^2 + 19x - 19}{x - 3} \times (x - 3) \equiv (x - 3)\left(3x^3 + 2x^2 - 4x + 7 + \frac{2}{x - 3} \right)$$

$$3x^4 - 7x^3 - 10x^2 + 19x - 19 \equiv (x - 3)(3x^3 + 2x^2 - 4x + 7) + (x - 3)\left(\frac{2}{x - 3} \right)$$

That is:

$$3x^4 - 7x^3 - 10x^2 + 19x - 19 \equiv (x - 3)(3x^3 + 2x^2 - 4x + 7) + 2$$

But this is the definition of a quotient and remainder. In other words,

$$f(x) \equiv (x - a)g(x) + R$$

'original polynomial $\equiv$ (divisor) $\times$ (quotient) + remainder'

The 'divisor' is the linear expression that the polynomial was divided by.

Let's try another example.

Divide $6x^5 + 2x^3 - 5x^2 + 2x - 1$ by $x - 1$.

$$
\begin{array}{r}
6x^4 + 6x^3 + 8x^2 + 3x + 5 \\
x - 1 \overline{)6x^5 + 0x^4 + 2x^3 - 5x^2 + 2x - 1} \\
\underline{6x^5 - 6x^4} \\
6x^4 + 2x^3 \\
\underline{6x^4 - 6x^3} \\
8x^3 - 5x^2 \\
\underline{8x^3 - 8x^2} \\
3x^2 + 2x \\
\underline{3x^2 - 3x} \\
5x - 1 \\
\underline{5x - 5} \\
4
\end{array}
$$

So:
$$\frac{6x^5 + 2x^3 - 5x^2 + 2x - 1}{x - 1} \equiv 6x^4 + 6x^3 + 8x^2 + 3x + 5 + \frac{4}{x - 1}$$

Multiply by $x - 1$:

$$\frac{6x^5 + 2x^3 - 5x^2 + 2x - 1}{x - 1} \times (x - 1) \equiv (x - 1)\left(6x^4 + 6x^3 + 8x^2 + 3x + 5 + \frac{4}{x - 1}\right)$$

$$6x^5 + 2x^3 - 5x^2 + 2x - 1 \equiv (x - 1)(6x^4 + 6x^3 + 8x^2 + 3x + 5) + (x - 1)\left(\frac{4}{x - 1}\right)$$

$$6x^5 + 2x^3 - 5x^2 + 2x - 1 \equiv (x - 1)(6x^4 + 6x^3 + 8x^2 + 3x + 5) + 4$$

Once again:

original polynomial $\equiv$ (divisor) $\times$ (quotient) + remainder

In the first example the divisor was $x - 3$. Look at the left-hand side of the identity:

$$3x^4 - 7x^3 - 10x^2 + 19x - 19 \equiv (x - 3)(3x^3 + 2x^2 - 4x + 7) + 2$$

and put $x = 3$.

$$
\begin{aligned}
\text{LHS} &= (3 \times 3^4) - (7 \times 3^3) - (10 \times 3^2) + (19 \times 3) - 19 \\
&= 243 - 189 - 90 + 57 - 19 \\
&= 300 - 298 \\
&= 2
\end{aligned}
$$

which is the remainder when the polynomial is divided by $x - 3$.

You will see the reason for this if you consider the right-hand side of the identity and put $x = 3$:

$$\begin{aligned}
\text{RHS} &= (x - 3)(3x^3 + 2x^2 - 4x + 7) + 2 \\
&= (3 - 3)(81 + 18 - 12 + 7) + 2 \\
&= (0 \times 94) + 2 \\
&= 0 + 2 \\
&= 2
\end{aligned}$$

That is, as soon as you put $x = 3$, the first bracket becomes zero, so the whole of the first expression becomes zero since anything multiplied by zero gives an answer of zero. So the right-hand side of the identity reduces to 'zero + the remainder of 2'.

You can do the same with the second example. Put $x = 1$ in the right-hand side of the identity:

$$6x^5 + 2x^3 - 5x^2 + 2x - 1 \equiv (x - 1)(6x^4 + 6x^3 + 8x^2 + 3x + 5) + 4$$
$$\text{LHS} = 6 + 2 - 5 + 2 - 1 = 4$$

which, once again, is the remainder. Again, you can see why this is so by considering the right-hand side of the identity with $x = 1$:

$$\begin{aligned}
\text{RHS} &= (1 - 1)(6 + 6 + 8 + 3 + 5) + 4 \\
&= (0 \times 28) + 4 \\
&= 4
\end{aligned}$$

Let us now consider the general situation. Suppose that when the polynomial $f(x)$, of degree n, where

$$f(x) \equiv a_n x^n + a_{n-1} x^{n-1} + \ldots + a_2 x^2 + a_1 x + a_0$$

is divided by $(x - \alpha)$, the quotient is a polynomial $g(x)$ (of degree $n - 1$) and there is a remainder R.

Then:
$$\frac{f(x)}{x - \alpha} \equiv g(x) + \frac{R}{x - \alpha}$$

or
$$\frac{f(x)}{\cancel{x - \alpha}} \times \cancel{(x - \alpha)} \equiv (x - \alpha) \times g(x) + \cancel{(x - \alpha)} \times \frac{R}{\cancel{x - \alpha}}$$
$$f(x) \equiv (x - \alpha)g(x) + R$$

Consider the LHS and RHS of this identity with $x = \alpha$.

$$\text{LHS} = f(\alpha)$$

$$\begin{aligned}
\text{RHS} &= (\alpha - \alpha) \times g(\alpha) + R \\
&= 0 \times g(\alpha) + R \\
&= R
\end{aligned}$$

So:
$$f(\alpha) = R$$

That is, if you substitute $x = \alpha$ into the polynomial you obtain the remainder that you would get if you divided f(x) by $(x - \alpha)$. This the **remainder theorem**. It can be stated more formally as:

■ **If a polynomial f(x) is divided by $x - \alpha$, the remainder is f(α).**

Example 9

Find the remainder when $3x^3 + 7x^2 + 2x + 1$ is divided by $x - 2$.

Let f$(x) \equiv 3x^3 + 7x^2 + 2x + 1$.

The remainder is

$$f(2) = (3 \times 8) + (7 \times 4) + (2 \times 2) + 1$$
$$= 24 + 28 + 4 + 1$$
$$= 57$$

Example 10

Find the remainder when $3x^4 - 2x^2 + 6x + 6$ is divided by $x + 1$.

Let f$(x) \equiv 3x^4 - 2x^2 + 6x + 6$.

The remainder is

$$f(\ 1) = 3 \quad 2 \quad 6 + 6$$
$$= 1$$

A more general case of the remainder theorem can be found by considering the remainder when the polynomial f(x) is divided by $\alpha x - \beta$. Suppose that the answer is, again, another polynomial g(x) and the remainder is R.

So:
$$\frac{f(x)}{\alpha x - \beta} \equiv g(x) + \frac{R}{\alpha x - \beta}$$

Multiply by $\alpha x - \beta$:

$$\frac{f(x)}{\alpha x - \beta} \times (\alpha x - \beta) \equiv (\alpha x - \beta) \times g(x) + (\alpha x - \beta) \times \frac{R}{\alpha x - \beta}$$
$$f(x) \equiv (\alpha x - \beta) \times g(x) + R$$

Now put $x = \dfrac{\beta}{\alpha}$:

$$f\left(\frac{\beta}{\alpha}\right) = \left(\alpha \frac{\beta}{\alpha} - \beta\right) \times g\left(\frac{\beta}{\alpha}\right) + R$$
$$f\left(\frac{\beta}{\alpha}\right) = 0 \times g\left(\frac{\beta}{\alpha}\right) + R$$

That is:
$$f\left(\frac{\beta}{\alpha}\right) = R$$

This is a more general form of the remainder theorem.

■ **If a polynomial f(x) is divided by $\alpha x - \beta$, the remainder is f$\left(\dfrac{\beta}{\alpha}\right)$.**

Example 11

Find the remainder when $2x^3 + 4x^2 - 6x + 1$ is divided by $2x - 1$.

Let $f(x) \equiv 2x^3 + 4x^2 - 6x + 1$.

Put $x = \frac{1}{2}$.

The remainder is:

$$
\begin{aligned}
f(\tfrac{1}{2}) &= (2 \times \tfrac{1}{8}) + (4 \times \tfrac{1}{4}) - (6 \times \tfrac{1}{2}) + 1 \\
&= \tfrac{1}{4} + 1 - 3 + 1 \\
&= -\tfrac{3}{4}
\end{aligned}
$$

Example 12

Find the remainder when $2x^2 + 3x - 1$ is divided by $3x + 2$.

Let $f(x) \equiv 2x^2 + 3x - 1$.

The remainder is:

$$
\begin{aligned}
f(-\tfrac{2}{3}) &= (2 \times \tfrac{4}{9}) + (3 \times -\tfrac{2}{3}) - 1 \\
&= \tfrac{8}{9} - 2 - 1 \\
&= -2\tfrac{1}{9}
\end{aligned}
$$

Example 13

When $2x^3 + ax^2 + x + 1$ is divided by $x + 2$ the remainder is -29. Find the value of the constant a.

Let $f(x) \equiv 2x^3 + ax^2 + x + 1$.

The remainder is:

$$
\begin{aligned}
f(-2) = -16 + 4a - 2 + 1 &= -29 \\
4a - 17 &= -29 \\
4a &= -12 \\
a &= -3
\end{aligned}
$$

If $x - \alpha$ is a factor of a polynomial $f(x)$ then $x - \alpha$ divides exactly into $f(x)$ and leaves no remainder. So:

$$
\frac{f(x)}{x - \alpha} \equiv g(x)
$$

That is:
$$
f(x) \equiv (x - \alpha) \times g(x)
$$

Put $x = \alpha$:
$$
\begin{aligned}
f(\alpha) &= (\alpha - \alpha) \times g(\alpha) \\
&= 0 \times g(\alpha) \\
&= 0
\end{aligned}
$$

This is the **factor theorem**, which you met in Book P1, chapter 1.

■ **If $x - \alpha$ is a factor of the polynomial $f(x)$, then $f(\alpha) = 0$.**

The factor theorem is a special case of the remainder theorem where the remainder is zero.

Example 14

Factorise completely $x^3 - 6x^2 + 11x - 6$.

Let $f(x) \equiv x^3 - 6x^2 + 11x - 6$.

Try $x = 1$: $\qquad\qquad$ $f(1) = 1 - 6 + 11 - 6 = 0$

So $x - 1$ is a factor of $f(x)$.

Try $x = 2$: $\qquad\qquad$ $f(2) = 8 - 24 + 22 - 6 = 0$

So $x - 2$ is a factor.

Try $x = 3$: $\qquad\qquad$ $f(3) = 27 - 54 + 33 - 6 = 0$

So $x - 3$ is a factor.

Thus: $\qquad$ $x^3 - 6x^2 + 11x - 6 \equiv (x - 1)(x - 2)(x - 3)$

Example 15

The polynomial $ax^3 - x^2 + bx + 6$ has a factor of $x + 2$, and when it is divided by $x + 1$ there is a remainder of 10. Find the values of the constants a and b. Find the values of x for which the polynomial is zero.

Let $f(x) \equiv ax^3 - x^2 + bx + 6$.

Since $(x + 2)$ is a factor of $f(x)$,
$$f(-2) = -8a - 4 - 2b + 6 = 0 \qquad\qquad (1)$$

Since there is a remainder of 10 when $f(x)$ is divided by $(x + 1)$,
$$f(-1) = -a - 1 - b + 6 = 10 \qquad\qquad (2)$$

From (1): $\qquad\qquad$ $-8a - 2b = -2$

From (2): $\qquad\qquad$ $-a - b = 5$

Solving these simultaneously:
$$4a + b = 1 \qquad\qquad (1)$$

$$a + b = -5 \qquad\qquad (2)$$

$(1) - (2)$: $\qquad\qquad$ $3a = 6$

$\qquad\qquad\qquad$ $a = 2$

Substitute in (2): $\qquad\qquad$ $2 + b = -5$

$\qquad\qquad\qquad$ $b = -7$

So: $\qquad\qquad$ $f(x) \equiv 2x^3 - x^2 - 7x + 6$

You know that $(x + 2)$ is a factor of $f(x)$.

Divide f(x) by $(x + 2)$:

$$
\begin{array}{r}
2x^2 - 5x + 3 \\
x + 2 \overline{)2x^3 - x^2 - 7x + 6} \\
\underline{2x^3 + 4x^2} \\
-5x^2 - 7x \\
\underline{-5x^2 - 10x} \\
3x + 6 \\
\underline{3x + 6}
\end{array}
$$

So :
$$
\begin{aligned}
2x^3 - x^2 - 7x + 6 &= (x + 2)(2x^2 - 5x + 3) \\
&= (x + 2)(2x - 3)(x - 1)
\end{aligned}
$$

When $2x^3 - x^2 - 7x + 6 = 0$,

$$(x + 2)(2x - 3)(x - 1) = 0$$

That is, $x = -2, \frac{3}{2}$ and 1.

Exercise 1B

Factorise:

1 $x^3 + 3x^2 - 4x - 12$
2 $x^3 + x^2 - 10x + 8$
3 $2x^3 + x^2 - 13x + 6$
4 $6x^3 - x^2 - 32x + 20$
5 $3x^3 + 17x^2 - 27x + 7$

Find the remainder when f(x) is divided by g(x):

6 f$(x) \equiv 3x^3 + 2x^2 - 6x + 1,\quad$ g$(x) \equiv x - 2$
7 f$(x) \equiv 5x^4 - 2x^3 + 3x - 2,\quad$ g$(x) \equiv x + 3$
8 f$(x) \equiv 2x^3 + 3x^2 - 7x - 14,\quad$ g$(x) \equiv x + 5$
9 f$(x) \equiv 4x^4 + 2x^2 - x - 7,\quad$ g$(x) \equiv 2x - 1$
10 f$(x) \equiv 4x^3 + 6x^2 + 3x + 2,\quad$ g$(x) \equiv 2x + 3$

11 When divided by $x + 1$, the polynomial $ax^3 - x^2 - x + 6$ leaves a remainder of 4. Find the value of the constant a.
12 When divided by $x - 2$, the polynomial $x^3 - ax^2 + 7x + 2$ leaves a remainder of -4. Find the value of the constant a.
13 When divided by $x + 3$, the polynomial $2x^3 + x^2 + ax + 1$ leaves a remainder of -53. Find the value of the constant a.

14 When divided by $3x - 1$, the polynomial $9x^3 + 9x^2 + ax + 2$ leaves a remainder of $4\frac{1}{3}$. Find the value of the constant a.

15 When divided by $2x - 3$, the polynomial $4x^3 - ax^2 - 2x + 7$ leaves a remainder of 13. Find the value of the constant a.

16 When divided by $x - 1$, the polynomial $ax^3 + x^2 + bx - 4$ leaves a remainder of -6. Given that $x - 2$ is a factor of the polynomial, find the values of the constants a and b.

17 When divided by $x + 1$, the polynomial $ax^3 + bx^2 - 13x + 6$ leaves a remainder of 18. Given that $2x - 1$ is a factor of the polynomial, find the values of the constants a and b.

18 When divided by $x - 1$, the polynomial $3x^3 + ax^2 - 5x + 2$ leaves a remainder of -4. Find the value of the constant a and hence factorise the polynomial.

19 Given that $2x^3 + ax^2 + x - 12$ leaves a remainder of 6 when divided by $x + 2$, find the value of the constant a. Hence solve the equation
$$2x^3 + ax^2 + x - 12 = 0$$

20 When divided by $x - 1$ the polynomial $ax^3 - 3x^2 + bx + 6$ leaves a remainder of -6. When divided by $x + 2$ it leaves a remainder of zero. Find the values of the constants a and b and hence solve the equation
$$ax^3 - 3x^2 + bx + 6 = 0$$

21 $$P(x) \equiv x^4 + x^3 - 5x^2 + ax + b$$
Given that $(x - 2)$ and $(x + 3)$ are factors of $P(x)$, find the values of the constants a and b.
Factorise $P(x)$ completely. [E]

22 $$f(x) \equiv 2x^3 + px^2 + qx + 6$$
where p and q are constants.
When $f(x)$ is divided by $(x + 1)$, the remainder is 12. When $f(x)$ is divided by $(x - 1)$, the remainder is -6.
(a) Find the value of p and the value of q.
(b) Show that $f(\frac{1}{2}) = 0$ and hence write $f(x)$ as the product of three linear factors. [E]

23 Find the value of the constant k so that the polynomial $P(x)$, where
$$P(x) \equiv x^2 + kx + 11$$
has a remainder 3 when it is divided by $(x - 2)$.
Show that, with this value of k, $P(x)$ is positive for all real x. [E]

24 The polynomial f(x), where

$$f(x) \equiv 2x^3 + Ax^2 + Bx - 3$$

is exactly divisible by $(x - 1)$ and has remainder $+9$ when divided by $(x + 2)$. Find the values of the constants A and B. Hence solve the equation f(x) = 0. **[E]**

25
$$f(x) \equiv 12x^3 + Ax^2 + Bx - 2$$
When f(x) is divided by $(x - 1)$ and $(x + 1)$ the remainders are 15 and -3 respectively.

(a) Find the values of the constants A and B.

(b) Using your values of A and B, find the values of x for which f(x) = 0.

1.3 The binomial series $(1 + x)^n$, $n \in \mathbb{Q}$, $|x| < 1$

Book P2 (page 46) introduces the binomial series $(1 + x)^n$, where n is a positive integer, as

$$(1 + x)^n = 1 + \binom{n}{1}x + \binom{n}{2}x^2 + \cdots + \binom{n}{r}x^r + \cdots + x^n$$

This may be written as:

$$(1 + x)^n = 1 + nx + \frac{n(n - 1)}{2!}x^2 + \frac{n(n - 1)(n - 2)}{3!}x^3 + \cdots + x^n$$

This result can be extended to include *all* values of n which are rational, such as $n = \frac{3}{4}$ or $n = -2$ or $n = -\frac{5}{4}$ and so on. You can write

$$(1 + x)^n = 1 + nx + \frac{n(n - 1)}{2!}x^2 + \frac{n(n - 1)(n - 2)}{3!}x^3 + \cdots + \frac{n(n - 1) \cdots (n - r + 1)}{r!}x^r + \cdots$$

For a positive integer value of n, this series is the one shown in Book P2 because the series comes to an end after $(n + 1)$ terms. If, on the other hand, n is *not* a positive integer, then the series is infinite and is only valid for $|x| < 1$; that is, $-1 < x < 1$ is an essential condition.

■ **In the form given, for $|x| < 1$,**

$$(1 + x)^n = 1 + nx + \frac{n(n - 1)}{2!}x^2 + \frac{n(n - 1)(n - 2)}{3!}x^3 + \cdots + \frac{n(n - 1) \cdots (n - r + 1)}{r!}x^r + \cdots$$

is called the binomial series of $(1 + x)^n$.

This result is quoted in the Edexcel formula book. You need to be able to expand series of the form $(a + bx)^n$ and to state the set of values of x for which the series is valid. The following examples are typical.

Example 16

Given that $|x| < 1$, expand in ascending powers of x up to and including the term in x^3: (a) $(1+x)^{-1}$ (b) $(1-x)^{-1}$ (c) $(1-x)^{-2}$.

In each case, use the expansion of $(1+x)^n$, where

$$(1+x)^n = 1 + nx + \frac{n(n-1)}{2!}x^2 + \frac{n(n-1)(n-2)}{3!}x^3 + \cdots$$

(a) You take x as x and $n = -1$ to obtain

$$(1+x)^{-1} = 1 + (-1)(x) + \frac{(-1)(-1-1)}{2!}x^2 + \frac{(-1)(-1-1)(-1-2)}{3!}x^3 + \cdots$$

$$= 1 - x + x^2 - x^3 + \cdots$$

(b) You take x as $-x$ and $n = -1$ to obtain

$$(1-x)^{-1} = 1 + (-1)(-x) + \frac{(-1)(-1-1)}{2!}(-x)^2 + \frac{(-1)(-1-1)(-1-2)}{3!}(-x)^3 + \cdots$$

$$= 1 + x + x^2 + x^3 + \cdots$$

(c) You take x as $-x$ and $n = -2$ to obtain

$$(1-x)^{-2} = 1 + (-2)(-x) + \frac{(-2)(-2-1)}{2!}(-x)^2 + \frac{(-2)(-2-1)(-2-2)}{3!}(-x)^3 + \cdots$$

$$= 1 + 2x + 3x^2 + 4x^3 + \cdots$$

Example 17

Expand (a) $(1+2x)^{\frac{3}{2}}$ (b) $(1-3x)^{-\frac{2}{3}}$ in ascending powers of x up to and including the term x^3. State the set of values of x for which your expansion is valid.

(a) In the general expansion of $(1+x)^n$, take x as $2x$ and take $n = \frac{3}{2}$ to obtain

$$(1+2x)^{\frac{3}{2}} = 1 + \frac{3}{2}(2x) + \frac{\frac{3}{2}(\frac{3}{2}-1)}{2!}(2x)^2 + \frac{\frac{3}{2}(\frac{3}{2}-1)(\frac{3}{2}-2)}{3!}(2x)^3 + \cdots$$

$$= 1 + 3x + \frac{3}{2}x^2 - \frac{1}{2}x^3 + \cdots$$

The series is valid for $|2x| < 1$, that is $|x| < \frac{1}{2}$ or $-\frac{1}{2} < x < \frac{1}{2}$.

(b) In the general expansion of $(1+x)^n$, take x as $-3x$ and let $n = -\frac{2}{3}$ to obtain:

$$(1-3x)^{-\frac{2}{3}} = 1 + (-\frac{2}{3})(-3x) + \frac{(-\frac{2}{3})(-\frac{2}{3}-1)}{2!}(-3x)^2 + \frac{(-\frac{2}{3})(-\frac{2}{3}-1)(-\frac{2}{3}-2)}{3!}(-3x)^3 + \cdots$$

$$= 1 + 2x + 5x^2 + \frac{40}{3}x^3 + \cdots$$

This series is valid for $|3x| < 1$, that is $|x| < \frac{1}{3}$ or $-\frac{1}{3} < x < \frac{1}{3}$.

You should note carefully how each series is built up, using brackets at the first stage to avoid introducing sign errors.

Example 18

Expand $(1 - 3x)^{\frac{1}{5}}$ in ascending powers of x up to and including the term in x^3. Using your series, take $x = \frac{1}{32}$ to find an approximation for $29^{\frac{1}{5}}$, giving your answer to five decimal places.

$$(1 - 3x)^{\frac{1}{5}} = 1 + \frac{1}{5}(-3x) + \frac{\frac{1}{5}(\frac{1}{5} - 1)}{2!}(-3x)^2 + \frac{\frac{1}{5}(\frac{1}{5} - 1)(\frac{1}{5} - 2)}{3!}(-3x)^3 + \cdots$$

$$= 1 - \frac{3}{5}x - \frac{18}{25}x^2 - \frac{162}{125}x^3 - \cdots$$

$$x = \frac{1}{32} \Rightarrow (1 - \frac{3}{32})^{\frac{1}{5}} = (\frac{29}{32})^{\frac{1}{5}} = \frac{1}{2}(29)^{\frac{1}{5}}$$

Putting $x = \frac{1}{32}$ in the series then gives:

$$\frac{1}{2}(29)^{\frac{1}{5}} \approx 1 - 0.018\,75 - 0.000\,703\,125 - 0.000\,039\,550\,781 - \cdots$$

$$\Rightarrow (29)^{\frac{1}{5}} = 1.961\,01 \text{ (to 5 d.p.)}$$

Example 19

Expand $(4 - x)^{-\frac{1}{2}}$ in ascending powers of x up to and including the term in x^3.

You should recognise at once that $(4 - x)^{-\frac{1}{2}}$ is *not* in the form $(1 + x)^n$. This means that you need to rewrite $(4 - x)^{-\frac{1}{2}}$ like this:

$$(4 - x)^{-\frac{1}{2}} = \left[4\left(1 - \frac{x}{4}\right)\right]^{-\frac{1}{2}} = 4^{-\frac{1}{2}}\left(1 - \frac{x}{4}\right)^{-\frac{1}{2}} = \frac{1}{2}\left(1 - \frac{x}{4}\right)^{-\frac{1}{2}}$$

and then $\left(1 - \frac{x}{4}\right)^{-\frac{1}{2}}$ can be expanded in the usual way, using the standard binomial series with x replaced by $-\frac{x}{4}$ and $n = -\frac{1}{2}$.

Notice also that the series is valid for $\left|\frac{x}{4}\right| < 1$, that is $|x| < 4$ or $-4 < x < 4$.

$$(4 - x)^{-\frac{1}{2}} = \frac{1}{2}\left(1 - \frac{x}{4}\right)^{-\frac{1}{2}}$$

$$= \frac{1}{2}\left[1 + (-\frac{1}{2})\left(-\frac{x}{4}\right) + \frac{(-\frac{1}{2})(-\frac{1}{2} - 1)}{2!}\left(-\frac{x}{4}\right)^2 + \frac{(-\frac{1}{2})(-\frac{1}{2} - 1)(-\frac{1}{2} - 2)}{3!}\left(-\frac{x}{4}\right)^3 + \cdots\right]$$

$$= \frac{1}{2}[1 + \frac{1}{8}x + \frac{3}{128}x^2 + \frac{5}{1024}x^3 + \cdots]$$

$$= \frac{1}{2} + \frac{1}{16}x + \frac{3}{256}x^2 + \frac{5}{2048}x^3 + \cdots$$

Example 20

$$f(x) \equiv \frac{x}{(3 - 2x)(2 - x)}$$

(a) Express $f(x)$ in partial fractions.

(b) Expand $f(x)$ in ascending powers of x up to and including the term in x^3.

(c) State the set of values of x for which the series is valid.

(a) Let
$$\frac{x}{(3-2x)(2-x)} \equiv \frac{A}{3-2x} + \frac{B}{2-x}$$

then:
$$x \equiv A(2-x) + B(3-2x)$$

When $x = 2$:
$$2 = B(3-4) \Rightarrow B = -2$$

When $x = \frac{3}{2}$:
$$\tfrac{3}{2} = A(2 - \tfrac{3}{2}) \Rightarrow A = 3$$

So:
$$f(x) \equiv \frac{3}{3-2x} - \frac{2}{2-x}$$

(b) Rewrite $f(x)$ as $3(3-2x)^{-1} - 2(2-x)^{-1}$
$$= 3(3^{-1})\left(1 - \frac{2x}{3}\right)^{-1} - 2(2^{-1})\left(1 - \frac{x}{2}\right)^{-1}$$
$$= \left(1 - \frac{2x}{3}\right)^{-1} - \left(1 - \frac{x}{2}\right)^{-1}$$

$$(1 - \tfrac{2}{3}x)^{-1} = 1 + (-1)(-\tfrac{2}{3}x) + \frac{(-1)(-2)}{2!}(-\tfrac{2}{3}x)^2 + \frac{(-1)(-2)(-3)}{3!}(-\tfrac{2}{3}x)^3 + \cdots$$
$$= 1 + \tfrac{2}{3}x + \tfrac{4}{9}x^2 + \tfrac{8}{27}x^3 + \cdots$$

$$(1 - \tfrac{1}{2}x)^{-1} = 1 + (-1)(-\tfrac{1}{2}x) + \frac{(-1)(-2)}{2!}(-\tfrac{1}{2}x)^2 + \frac{(-1)(-2)(-3)}{3!}(-\tfrac{1}{2}x)^3 + \cdots$$
$$= 1 + \tfrac{1}{2}x + \tfrac{1}{4}x^2 + \tfrac{1}{8}x^3 + \cdots$$

So:
$$f(x) = (\tfrac{2}{3} - \tfrac{1}{2})x + (\tfrac{4}{9} - \tfrac{1}{4})x^2 + (\tfrac{8}{27} - \tfrac{1}{8})x^3 + \cdots$$
$$= \tfrac{1}{6}x + \tfrac{7}{36}x^2 + \tfrac{37}{216}x^3 + \cdots$$

(c) The expansion of $(1 - \tfrac{2}{3}x)^{-1}$ is valid for $|\tfrac{2}{3}x| < 1$, that is:
$$-\tfrac{3}{2} < x < \tfrac{3}{2}$$
The expansion of $(1 - \tfrac{1}{2}x)^{-1}$ is valid for $|\tfrac{1}{2}x| < 1$, that is:
$$-2 < x < 2$$

The expansion of $f(x)$ is valid when *both* sets of conditions are satisfied, that is, for the *intersection* of these two sets, when:
$$-\tfrac{3}{2} < x < \tfrac{3}{2}$$

Exercise 1C

In questions 1–15, find, in ascending powers of x, the expansions up to and including the term in x^3, simplifying the coefficients. State the set of values of x for which each expansion is valid.

1 $(1+x)^{-2}$ **2** $(1-x)^{-3}$ **3** $(1-x)^{-5}$

4 $(1+x)^{-\frac{1}{2}}$ **5** $(1+x)^{\frac{3}{2}}$ **6** $(1-x)^{\frac{3}{4}}$

7 $(1-3x)^{\frac{1}{3}}$ **8** $(1+3x)^{-\frac{1}{3}}$ **9** $(1-\tfrac{1}{2}x)^{-2}$

10 $(1+6x)^{-1}$ **11** $(3+x)^{-1}$ **12** $(2-x)^{-2}$

13 $(4+3x)^{\frac{1}{2}}$ **14** $(8-5x)^{\frac{1}{3}}$ **15** $(100+x)^{-\frac{1}{2}}$

By using partial fractions find, in ascending powers of x, up to and including the term in x^3, expansions for the functions of x given in questions 16–20. State the set of values of x for which the expansion is valid.

16 $\dfrac{2 - 3x}{1 - 3x + 2x^2}$ **17** $\dfrac{3}{1 + x - 2x^2}$ **18** $\dfrac{2}{x^2 + 2x - 8}$

19 $\dfrac{1}{x^2 + 3x + 2}$ **20** $\dfrac{8 - x}{x^2 - x - 6}$

21 Given that $|x| < \frac{1}{2}$, expand $(1 + x)^2 (1 - 2x)^{-\frac{1}{2}}$ in ascending powers of x up to and including the term in x^3, simplifying each coefficient.

22 Given that $|x| > 2$ find the first four terms in the series expansion of $\left(1 - \dfrac{2}{x}\right)^{\frac{1}{2}}$ in descending powers of x.

By taking $x = 200$ use your series to find a value of $\sqrt{99}$, giving your answer to 7 decimal places. Use your series to find $\sqrt{101}$ to the same degree of accuracy.

23 The series expansion of $(1 + px)^q$ in ascending powers of x has coefficients of -10 and 75 in the x and x^2 terms respectively.
(a) Find the value of p and of q.
(b) Find the coefficients of the x^3 and x^4 terms in the expansion.
(c) State the set of values of x for which the series is valid.

24 Given that $|x| < 1$, expand $\left(\dfrac{1 + x}{1 - x}\right)^{\frac{1}{3}}$ in ascending powers of x up to and including the term in x^2.

25 The coefficients of x and x^2 in the expansion of $(1 + px + qx^2)^{-2}$ in ascending powers of x are 4 and 14 respectively. Find the values of p and q.

26 The coefficients of the x and x^2 terms in the expansion of $(1 + px)^q$ in ascending powers of x are -6 and 6 respectively.
(a) Find the value of p and of q.
(b) Find the x^3 term and the x^4 term in the expansion.
(c) State the set of values of x for which the expansion is valid.

SUMMARY OF KEY POINTS

1 A proper fraction with linear factors in the denominator, such as

$$\frac{f(x)}{(x-1)(x-2)(x-3)}$$

has partial fractions of the form

$$\frac{A}{x-1}+\frac{B}{x-2}+\frac{C}{x-3}$$

where A, B and C are constants.

2 A proper fraction with a non-reducible quadratic factor in the denominator, such as

$$\frac{f(x)}{(x+2)(x^2+2x+3)}$$

has partial fractions of the form

$$\frac{A}{x+2}+\frac{Bx+C}{x^2+2x+3}$$

where A, B and C are constants.

3 A proper fraction with a repeated factor in the denominator, such as

$$\frac{f(x)}{(x+2)^3}$$

has partial fractions of the form

$$\frac{A}{x+2}+\frac{B}{(x+2)^2}+\frac{C}{(x+2)^3}$$

where A, B and C are constants.

4 An improper fraction is one where the degree of the numerator is greater than or equal to the degree of the denominator.

5 With an improper fraction the denominator must first be divided into the numerator before attempting to split it into partial fractions.

6 The remainder theorem states that if a polynomial $f(x)$ is divided by $\alpha x-\beta$, the remainder is $f\left(\dfrac{\beta}{\alpha}\right)$.

7 The factor theorem states that if $x-\alpha$ is a factor of the polynomial $f(x)$, then $f(\alpha)=0$. Also, if $f(\alpha)=0$, then $x-\alpha$ is a factor of the polynomial $f(x)$.

8   The binomial series

$$(1+x)^n = 1 + nx + \frac{n(n-1)}{2!}x^2 + \frac{n(n-1)(n-2)}{3!}x^3 + \cdots$$
$$+ \frac{n(n-1)(n-2)\cdots(n-r+1)}{r!}x^r + \cdots$$

(a) has $(n+1)$ terms only if n is a positive integer

(b) has an infinite number of terms if n is rational, but not a positive integer.

The series expansion is only valid in this case if $|x| < 1$.

Differentiation

2

2.1 Composite functions

In chapter 2 of Book P2 you learned how to form and to evaluate composite functions.

Example 1

Given that $f(x) \equiv x^2$ and $g(x) \equiv 3x - 1$, then the composite function fg is given by

$$fg(x) \equiv f(3x - 1) \equiv (3x - 1)^2$$

The composite function gf is given by

$$gf(x) \equiv g(x^2) \equiv 3x^2 - 1$$

You have now reached the stage where you will need to differentiate composite functions. In Books P1 and P2, you learned these important results:

y or $f(x)$	$\dfrac{\mathrm{d}y}{\mathrm{d}x}$ or $f'(x)$
x^n (n rational)	nx^{n-1}
e^x	e^x
$\ln x$	$\dfrac{1}{x}$
$u \pm v$	$\dfrac{\mathrm{d}u}{\mathrm{d}x} \pm \dfrac{\mathrm{d}v}{\mathrm{d}x}$

Consider how you could differentiate $(3x - 1)^2$ using your knowledge from Book P1.

First write $\qquad y = (3x - 1)^2$

Expand to give $\qquad y = 9x^2 - 6x + 1$

Differentiate: $\qquad \dfrac{\mathrm{d}y}{\mathrm{d}x} = 18x - 6 = 6(3x - 1)$

This approach is laborious, and it gets worse! Consider, for example, differentiating $(3x - 1)^6$ in this way. Especially when the actual answer, $18(3x - 1)^5$, can be obtained so simply by the method you are about to learn.

2.2 Differentiating composite functions using the chain rule

Suppose that y is a function of t and that t is a function of x. The idea of taking an increment (small change) was introduced in Book P1. Now suppose that a small change δx in the variable x gives rise to small changes δy and δt in the variables y and t respectively.

Also
$$\frac{\delta y}{\delta x} = \frac{\delta y}{\delta t} \cdot \frac{\delta t}{\delta x}$$

As δy, δt and $\delta x \to 0$ we may assume that

$$\frac{\delta y}{\delta x} \to \frac{\mathrm{d}y}{\mathrm{d}x}, \; \frac{\delta y}{\delta t} \to \frac{\mathrm{d}y}{\mathrm{d}t} \; \text{ and } \; \frac{\delta t}{\delta x} \to \frac{\mathrm{d}t}{\mathrm{d}x}$$

This leads to the important result

■
$$\frac{\mathrm{d}y}{\mathrm{d}x} = \frac{\mathrm{d}y}{\mathrm{d}t} \cdot \frac{\mathrm{d}t}{\mathrm{d}x}$$

This is called the **chain rule**. You will find that almost every differentiation you undertake will involve some use of the chain rule. You need to know this result, but you will not be expected to prove it.

Knowing how to use the chain rule is of prime importance.

Example 2
Differentiate (a) $(3x-1)^2$ (b) $(3x-1)^6$ with respect to x.

(a) Write $y = (3x-1)^2$ and $t = 3x-1$

Then: $\qquad y = t^2$ and $t = 3x-1$

Differentiating gives:

$$\frac{\mathrm{d}y}{\mathrm{d}t} = 2t \quad \text{and} \quad \frac{\mathrm{d}t}{\mathrm{d}x} = 3$$

Using the chain rule gives

$$\frac{\mathrm{d}y}{\mathrm{d}x} = \frac{\mathrm{d}y}{\mathrm{d}t} \cdot \frac{\mathrm{d}t}{\mathrm{d}x} = 2t \times 3 = 6t$$

But $t = 3x-1$, so

$$\frac{\mathrm{d}y}{\mathrm{d}x} = 6(3x-1)$$

(b) Write $y = (3x-1)^6$ and $t = 3x-1$

Then: $\qquad y = t^6$ and $t = 3x-1$

Differentiating: $\qquad \dfrac{\mathrm{d}y}{\mathrm{d}t} = 6t^5$ and $\dfrac{\mathrm{d}t}{\mathrm{d}x} = 3$

Using the chain rule:

$$\frac{dy}{dx} = \frac{dy}{dt} \cdot \frac{dt}{dx} = 6t^5 \times 3 = 18t^5$$

But $t = 3x - 1$; so

$$\frac{dy}{dx} = 18(3x - 1)^5$$

Example 3

Find $\dfrac{dy}{dx}$, where $y = \ln(x^2 - 3x + 5)$.

Write $t = x^2 - 3x + 5$; then:

$$y = \ln t \quad \text{and} \quad t = x^2 - 3x + 5$$

Differentiating: $\qquad \dfrac{dy}{dt} = \dfrac{1}{t}$ and $\dfrac{dt}{dx} = 2x - 3$

Using the chain rule:

$$\frac{dy}{dx} = \frac{dy}{dt} \cdot \frac{dt}{dx} = \frac{1}{t} \times (2x - 3)$$

But $t = x^2 - 3x + 5$; so:

$$\frac{dy}{dx} = \frac{2x - 3}{x^2 - 3x + 5}$$

Example 4

Find $\dfrac{dy}{dx}$, where $y = e^{\sqrt{x}}$.

Write $t = \sqrt{x} = x^{\frac{1}{2}}$; then:

$$y = e^t$$

Differentiating: $\qquad \dfrac{dt}{dx} = \tfrac{1}{2}x^{-\frac{1}{2}} = \dfrac{1}{2\sqrt{x}}$

and $\qquad \dfrac{dy}{dt} = e^t$

Using the chain rule:

$$\frac{dy}{dx} = \frac{dy}{dt} \cdot \frac{dt}{dx} = e^t \times \frac{1}{2\sqrt{x}}$$

But $t = \sqrt{x}$; so:

$$\frac{dy}{dx} = \frac{1}{2\sqrt{x}} e^{\sqrt{x}}$$

Example 5

Find the gradient of the curve with equation $y = \dfrac{1}{(3x+2)^2}$ at the point where $x = \frac{1}{3}$.

Write $t = 3x + 2$ and then $y = t^{-2}$.

Differentiating: $\qquad \dfrac{dt}{dx} = 3$ and $\dfrac{dy}{dt} = -2t^{-3}$

Using the chain rule:

$$\frac{dy}{dx} = \frac{dy}{dt} \cdot \frac{dt}{dx} = -6t^{-3}$$

But $t = 3x + 2$; so:

$$\frac{dy}{dx} = -6(3x+2)^{-3} = \frac{-6}{(3x+2)^3}$$

The gradient of the curve with equation $y = \dfrac{1}{(3x+2)^2}$ at the point where $x = \frac{1}{3}$ is found by putting $x = \frac{1}{3}$ in the expression for $\dfrac{dy}{dx}$.

That is, at $x = \frac{1}{3}$,

$$\frac{dy}{dx} = \frac{-6}{(1+2)^3} = \frac{-6}{27} = -\frac{2}{9}$$

Exercise 2A

Differentiate with respect to x:

1 $(x+2)^2$ **2** $(x-5)^7$ **3** $(3-x)^4$ **4** $(2x-1)^3$

5 $(3x-7)^4$ **6** $(x-4)^{-2}$ **7** $(3-x)^{-3}$ **8** $\dfrac{1}{x-7}$

9 $\dfrac{3}{5-x}$ **10** $(6x^2+1)^4$ **11** $\ln(x-2)$ **12** e^{x^2}

13 e^{3-x} **14** $\ln(6-x^2)$ **15** $\dfrac{1}{x^2-7x+2}$ **16** $\dfrac{-6}{3-4x^3}$

17 $\sqrt{(5+4x)}$ **18** $(1-x^{\frac{1}{2}})^4$ **19** $(x^{\frac{2}{3}}+5)^{-3}$ **20** e^{x^2-3x}

21 $\left(1+\dfrac{1}{x}\right)^4$ **22** $\left(x^2-\dfrac{1}{x}\right)^{-3}$ **23** $(1-x^2)^{\frac{1}{2}}$ **24** $\dfrac{1}{(1+x^2)^2}$

25 $\ln(3x^2-4x+3)$ **26** $\ln(x^{\frac{1}{2}}+x^{-\frac{1}{2}})$ **27** $\ln(e^x+5)$ **28** $e^{\ln x}$

29 $\dfrac{1}{(6x^3-5)^3}$ **30** $\dfrac{1}{\ln x}$

31 Given that $y = \dfrac{1}{(1+\sqrt{x})^3}$, find the value of $\dfrac{dy}{dx}$ at $x = 4$.

32 Given that $y = \ln(x^3 + 4x)$, find the value of $\dfrac{dy}{dx}$ at $x = 1$.

33 Given that $y = e^{x^2 - x}$, show that the value of $\dfrac{dy}{dx}$ is 1 if $x = 1$ and $\dfrac{dy}{dx} = -1$ if $x = 0$.

34 Given that $y = (7x^2 - 1)^{\frac{1}{3}}$, find the value of $\dfrac{dy}{dx}$ at $x = 2$.

35 Find $\dfrac{d}{dx}(2x^3 - 5x)^{-4}$ at $x = -1$.

2.3 Differentiating products

You know already how to differentiate sums and differences of two functions, u and v, of x using the formula

$$\frac{d}{dx}(u \pm v) = \frac{du}{dx} \pm \frac{dv}{dx}$$

Now you will learn how to differentiate uv, where u and v are functions of x.

Consider $y = uv$, where u and v are functions of x.

Suppose that you make a small change, δx, in x and this in turn gives rise to small changes δy, δu and δv in y, u and v respectively.

Then:
$$y + \delta y = (u + \delta u)(v + \delta v)$$
$$= uv + u\delta v + v\delta u + \delta u \delta v$$

or:
$$\delta y = uv + u\delta v + v\delta u + \delta u \delta v - y$$

But $y = uv$, so
$$\delta y = u\delta v + v\delta u + \delta u \delta v$$

So:
$$\frac{\delta y}{\delta x} = u\frac{\delta v}{\delta x} + v\frac{\delta u}{\delta x} + \frac{\delta u}{\delta x}\delta v$$

As $\delta x \to 0$, $\quad \dfrac{\delta u}{\delta x}\delta v = \dfrac{\delta u}{\delta x} \cdot \dfrac{\delta v}{\delta x} \cdot \delta x \to 0$

and $\dfrac{\delta y}{\delta x} \to \dfrac{dy}{dx}$, $\dfrac{\delta u}{\delta x} \to \dfrac{du}{dx}$ and $\dfrac{\delta v}{\delta x} \to \dfrac{dv}{dx}$

So:

$$\blacksquare \qquad \frac{dy}{dx} = v\frac{du}{dx} + u\frac{dv}{dx}$$

This is known as the **product rule**. You should learn it because you will often need to use it. You are not expected to know how to prove this formula in examinations.

Example 6
Given that $y = 3x^3(2x - 5)^4$, find $\dfrac{dy}{dx}$.

Put $u = 3x^3$; so $\dfrac{du}{dx} = 9x^2$

Put $v = (2x - 5)^4$; then by the chain rule:

$$\frac{dv}{dx} = 4(2x - 5)^3(2) = 8(2x - 5)^3$$

$$\begin{aligned}
\frac{dy}{dx} &= v\frac{du}{dx} + u\frac{dv}{dx} \\
&= 9x^2(2x - 5)^4 + 3x^3 \cdot 8(2x - 5)^3 \\
&= 3x^2(2x - 5)^3[3(2x - 5) + 8x] \\
&= 3x^2(2x - 5)^3[6x - 15 + 8x] \\
&= 3x^2(2x - 5)^3(14x - 15)
\end{aligned}$$

Notice that you will usually need to 'tidy up' your answer after applying the product formula.

Example 7
Given that $y = e^{-2x}\sqrt{(x^2 + 1)}$, find $\dfrac{dy}{dx}$.

Write $u = e^{-2x}$; then $\dfrac{du}{dx} = -2e^{-2x}$

Put $v = \sqrt{(x^2 + 1)} = (x^2 + 1)^{\frac{1}{2}}$

By the chain rule: $\dfrac{dv}{dx} = \tfrac{1}{2}(x^2 + 1)^{-\frac{1}{2}}(2x)$

$$= \frac{x}{\sqrt{(x^2 + 1)}}$$

$$\begin{aligned}
\frac{dy}{dx} &= v\frac{du}{dx} + u\frac{dv}{dx} \\
&= (-2e^{-2x})\sqrt{(x^2 + 1)} + e^{-2x} \cdot \frac{x}{\sqrt{(x^2 + 1)}}
\end{aligned}$$

Take out the common factor $\dfrac{e^{-2x}}{\sqrt{(x^2 + 1)}}$ to give

$$\begin{aligned}
\frac{dy}{dx} &= \frac{e^{-2x}}{\sqrt{(x^2 + 1)}}[x - 2\sqrt{(x^2 + 1)}\sqrt{(x^2 + 1)}] \\
&= \frac{e^{-2x}}{\sqrt{(x^2 + 1)}}(x - 2x^2 - 2)
\end{aligned}$$

2.4 Differentiating quotients

You can write the quotient $\dfrac{u}{v}$, where u and v are functions of x, as:

$$y = \frac{u}{v} = uv^{-1}$$

By the product rule then,

$$\frac{dy}{dx} = v^{-1}\frac{du}{dx} + u\frac{d}{dx}(v^{-1})$$

As v is a function of x, you can write $t = v^{-1}$.

Then:
$$\frac{dt}{dv} = -v^{-2}$$

By the chain rule,

$$\frac{dt}{dx} = \frac{dt}{dv}\cdot\frac{dv}{dx} = -v^{-2}\frac{dv}{dx}$$

So:
$$\frac{dy}{dx} = v^{-1}\frac{du}{dx} - uv^{-2}\frac{dv}{dx}$$

$$= \frac{1}{v}\frac{du}{dx} - \frac{u}{v^2}\frac{dv}{dx}$$

$$\blacksquare \qquad \frac{dy}{dx} = \frac{v\dfrac{du}{dx} - u\dfrac{dv}{dx}}{v^2}$$

This result is known as the **quotient rule**. Here are some examples of how to use it.

Example 8

Given that $y = \dfrac{x^2 - 1}{x^2 + 1}$, find $\dfrac{dy}{dx}$.

Put $u = x^2 - 1$, then:
$$\frac{du}{dx} = 2x$$

Put $v = x^2 + 1$, then:
$$\frac{dv}{dx} = 2x$$

Apply the quotient formula:

$$\frac{dy}{dx} = \frac{v\dfrac{du}{dx} - u\dfrac{dv}{dx}}{v^2}$$

$$\frac{dy}{dx} = \frac{2x(x^2 + 1) - 2x(x^2 - 1)}{(x^2 + 1)^2}$$

$$= \frac{2x^3 + 2x - 2x^3 + 2x}{(x^2 + 1)^2}$$

So:
$$\frac{dy}{dx} = \frac{4x}{(x^2 + 1)^2}$$

Alternatively, you could write $y = (x^2 - 1)(x^2 + 1)^{-1}$ and use the product rule. Then:

$$u = x^2 - 1 \qquad \frac{du}{dx} = 2x$$

$v = (x^2 + 1)^{-1}$ and, by the chain rule, $\dfrac{dv}{dx} = -2x(x^2 + 1)^{-2}$

$$\begin{aligned}
\frac{dy}{dx} &= v\frac{du}{dx} + u\frac{dv}{dx} \\
&= 2x(x^2 + 1)^{-1} + (x^2 - 1)[-2x(x^2 + 1)^{-2}] \\
&= \frac{2x}{x^2 + 1} - \frac{2x(x^2 - 1)}{(x^2 + 1)^2} \\
&= \frac{2x(x^2 + 1) - 2x(x^2 - 1)}{(x^2 + 1)^2} \\
&= \frac{2x^3 + 2x - 2x^3 + 2x}{(x^2 + 1)^2} \\
&= \frac{4x}{(x^2 + 1)^2}
\end{aligned}$$

As you can see, the quotient rule formula reduces the amount of algebraic processing you have to do and this is why the first method is recommended.

Example 9

Given that $y = \dfrac{\sqrt{(3 - 4x)}}{\ln(2x - 3)}$, find $\dfrac{dy}{dx}$

Put $u = \sqrt{(3 - 4x)} = (3 - 4x)^{\frac{1}{2}}$; then: $\dfrac{du}{dx} = \frac{1}{2}(3 - 4x)^{-\frac{1}{2}}(-4)$
$$= -2(3 - 4x)^{-\frac{1}{2}}$$

Put $v = \ln(2x - 3)$; then: $\dfrac{dv}{dx} = \dfrac{2}{2x - 3}$

using the chain rule in each case.

So:

$$\begin{aligned}
\frac{dy}{dx} &= \frac{v\dfrac{du}{dx} - u\dfrac{dv}{dx}}{v^2} \\
&= \frac{-2(3 - 4x)^{-\frac{1}{2}}\ln(2x - 3) - \dfrac{2(3 - 4x)^{\frac{1}{2}}}{2x - 3}}{[\ln(2x - 3)]^2} \\
&= \frac{\dfrac{-2\ln(2x - 3)}{\sqrt{(3 - 4x)}} - \dfrac{2\sqrt{(3 - 4x)}}{2x - 3}}{[\ln(2x - 3)]^2} \\
&= \frac{-2(2x - 3)\ln(2x - 3) - 2(3 - 4x)}{(2x - 3)\sqrt{(3 - 4x)}[\ln(2x - 3)]^2}
\end{aligned}$$

Exercise 2B

Differentiate each of the following with respect to x:

1 $x^3(x-1)^2$
2 $x^4(2x+1)^3$
3 $x^2 e^x$
4 $(2x-1)\ln x$
5 $e^{2x}\ln x$
6 $(x+1)\ln(3x-1)$
7 $x(2x-1)^5$
8 $(2x-3)^3(x^2+1)^2$
9 $x(x^2+1)^4$
10 $(2x-3)\sqrt{(x^2-1)}$
11 $(x-1)^2\sqrt{x}$
12 $(x-2)^{\frac{2}{3}}(x+2)^{\frac{3}{4}}$
13 $(2x^2-1)\sqrt{(4x-3)}$
14 $x^{\frac{1}{2}}\ln(x^2-1)$
15 $(x-1)^3 e^{x^2-2x}$
16 $\dfrac{x}{x-1}$

17 $\dfrac{x^2}{1-x}$
18 $\dfrac{x^3}{2-x^2}$
19 $\dfrac{e^x}{2x+1}$

20 $\dfrac{\ln(1-x^2)}{x^3}$
21 $\dfrac{x^3-1}{x^3+1}$
22 $\dfrac{2x}{\sqrt{(x-1)}}$

23 $\dfrac{2x-1}{x^2-1}$
24 $\dfrac{\sqrt{x}}{x^4+1}$
25 $\dfrac{3e^x-1}{3e^x+1}$

26 $\dfrac{e^{x^2}}{\ln(2x+1)}$
27 $\dfrac{\ln(x^2+1)}{x^2-1}$
28 $\dfrac{x^2(x-\frac{1}{x})}{x+\frac{1}{x}}$

29 $\dfrac{\sqrt{(4x^3-1)}}{\ln(2x-3)}$
30 $\dfrac{\ln(x^2+1)}{e^x+1}$

31 Given that $y=\dfrac{x}{x^2+1}$, find the values of x for which $\dfrac{dy}{dx}=0$.

32 Given that $y=(1+x^2)^{\frac{1}{2}}$, find an expression for $\dfrac{dy}{dx}$. Hence find the value of $\dfrac{dy}{dx}$ when $x=0$ and when $x=\frac{3}{4}$.

33 Given that $y=x^3(x-3)^2$, find the values of x and y when $\dfrac{dy}{dx}=0$.

34 Given that $y=\dfrac{x^{\frac{1}{2}}-1}{x^{\frac{1}{2}}+1}$, find the value of $\dfrac{dy}{dx}$ when $x=9$.

35 Given that $f(x)=x\ln(x^3-4)$, find the value of $f'(2)$.

2.5 Connected rates of change

You learned in Book P1 that the derivative is used to measure rates of change. By using the chain rule you can find rates of change that are related to other rates of change, as the following examples illustrate. Sometimes you may also need to use the product or quotient rule.

Example 10

The radius of a spherical balloon is increasing at the rate of $0.2\,\mathrm{m\,s^{-1}}$. Find the rate of increase of (a) the volume (b) the surface area of the balloon at the instant when the radius is $1.6\,\mathrm{m}$.

(a) The volume $V\,\mathrm{m^3}$ of a sphere, radius $r\,\mathrm{m}$, is given by the formula $V = \frac{4}{3}\pi r^3$.

We know that $\dfrac{\mathrm{d}r}{\mathrm{d}t} = 0.2$ and also, by differentiating, that:

$$\frac{\mathrm{d}V}{\mathrm{d}r} = 4\pi r^2$$

Use the chain rule:

$$\frac{\mathrm{d}V}{\mathrm{d}t} = \frac{\mathrm{d}V}{\mathrm{d}r} \cdot \frac{\mathrm{d}r}{\mathrm{d}t}$$

When $r = 1.6$, the rate of change of V with respect to t is

$$\frac{\mathrm{d}V}{\mathrm{d}t} = 4\pi (1.6)^2 (0.2)\ \mathrm{m^3\,s^{-1}}$$

$$= 6.43\ \mathrm{m^3\,s^{-1}}$$

(b) The surface area $A\,\mathrm{m^2}$ of a sphere, radius $r\,\mathrm{m}$, is given by the formula $A = 4\pi r^2$.

We know that $\dfrac{\mathrm{d}r}{\mathrm{d}t} = 0.2$ and also, by differentiating, that:

$$\frac{\mathrm{d}A}{\mathrm{d}r} = 8\pi r$$

Use the chain rule:

$$\frac{\mathrm{d}A}{\mathrm{d}t} = \frac{\mathrm{d}A}{\mathrm{d}r} \cdot \frac{\mathrm{d}r}{\mathrm{d}t}$$

When $r = 1.6$, the rate of change of A with respect to t is

$$\frac{\mathrm{d}A}{\mathrm{d}t} = 8\pi (1.6)(0.2)\ \mathrm{m^2\,s^{-1}}$$

$$= 8.04\ \mathrm{m^2\,s^{-1}}$$

Example 11

Given that $P = x(x^2 + 4)^{\frac{1}{2}}$, find $\dfrac{\mathrm{d}P}{\mathrm{d}t}$ when $x = 2$ and $\dfrac{\mathrm{d}x}{\mathrm{d}t} = 3$.

First you need to find $\dfrac{\mathrm{d}P}{\mathrm{d}x}$ using the product rule, so write:

$$u = x \quad \text{and} \quad v = (x^2 + 4)^{\frac{1}{2}}$$

Then:

$$\frac{du}{dx} = 1 \quad \text{and} \quad \frac{dv}{dx} = \tfrac{1}{2}(x^2 + 4)^{-\frac{1}{2}}(2x)$$

$$= x(x^2 + 4)^{-\frac{1}{2}}$$

$$\frac{dP}{dx} = v\frac{du}{dx} + u\frac{dv}{dx}$$

$$= (x^2 + 4)^{\frac{1}{2}} + x^2(x^2 + 4)^{-\frac{1}{2}}$$

At $x = 2$, $\qquad \dfrac{dP}{dx} = (2^2 + 4)^{\frac{1}{2}} + 2^2(2^2 + 4)^{-\frac{1}{2}}$

$$= 8^{\frac{1}{2}} + \frac{4}{8^{\frac{1}{2}}}$$

$$= \frac{8 + 4}{8^{\frac{1}{2}}} = \frac{12}{2 \times 2^{\frac{1}{2}}} = \frac{6}{2^{\frac{1}{2}}} \times \frac{2^{\frac{1}{2}}}{2^{\frac{1}{2}}} = 3 \times 2^{\frac{1}{2}}$$

$$\frac{dP}{dt} = \frac{dP}{dx} \cdot \frac{dx}{dt} \quad \text{(chain rule)}$$

$$= 3 \times 2^{\frac{1}{2}} \times 3$$

$$= 9 \times 2^{\frac{1}{2}} \text{ or } 9\sqrt{2}$$

2.6 Further examples using differentiation

Example 12

Express $\dfrac{7x + 1}{(1 - 2x)(1 + x)}$ in partial fractions and hence find the

values of $\dfrac{dy}{dx}$ and $\dfrac{d^2y}{dx^2}$, where $y = \dfrac{7x + 1}{(1 - 2x)(1 + x)}$, at $x = 0$.

Using the work from chapter 1, you can write

$$\frac{7x + 1}{(1 - 2x)(1 + x)} \equiv \frac{A}{1 - 2x} + \frac{B}{1 + x}$$

So $7x + 1 \equiv A(1 + x) + B(1 - 2x)$ is true for all values of x.

$x = -1$: $\qquad -7 + 1 = 0 + B(1 + 2) \Rightarrow B = -2$

$x = \tfrac{1}{2}$: $\qquad \tfrac{7}{2} + 1 = A(1 + \tfrac{1}{2}) + 0 \Rightarrow A = 3$

Hence: $\qquad y = \dfrac{7x + 1}{(1 - 2x)(1 + x)} \equiv \dfrac{3}{(1 - 2x)} - \dfrac{2}{1 + x}$

$$\equiv 3(1 - 2x)^{-1} - 2(1 + x)^{-1}$$

Using the chain rule of differentiation:

$$\frac{dy}{dx} = 3(-2)(-1)(1-2x)^{-2} - 2(-1)(1+x)^{-2}$$

$$= \frac{6}{(1-2x)^2} + \frac{2}{(1+x)^2}$$

At $x = 0$:
$$\frac{dy}{dx} = \frac{6}{1} + \frac{2}{1} = 8$$

$$\frac{d^2y}{dx^2} = \frac{d}{dx}\left(\frac{dy}{dx}\right) = \frac{d}{dx}\left[\frac{6}{(1-2x)^2} + \frac{2}{(1+x)^2}\right]$$

$$\frac{d^2y}{dx^2} = \frac{d}{dx}[6(1-2x)^{-2} + 2(1+x)^{-2}]$$

Using the chain rule:

$$\frac{d^2y}{dx^2} = 6(-2)(-2)(1-2x)^{-3} + 2(-2)(1+x)^{-3}$$

$$= \frac{24}{(1-2x)^3} - \frac{4}{(1+x)^3}$$

At $x = 0$:
$$\frac{d^2y}{dx^2} = \frac{24}{1^3} - \frac{4}{1^3} = 20$$

Example 13

Find the turning points and the points of inflexion on the curve with equation $y = x^2 e^x$.

Using the product rule:
$$u = x^2 \qquad v = e^x$$

$$\frac{du}{dx} = 2x \qquad \frac{dv}{dx} = e^x$$

$$\frac{dy}{dx} = v\frac{du}{dx} + u\frac{dv}{dx} = 2xe^x + x^2 e^x$$

$$= xe^x(x+2)$$

For turning points $\frac{dy}{dx} = 0$.

$$xe^x(x+2) = 0 \Rightarrow x = 0 \quad \text{or} \quad x = -2$$

Remember that $e^x > 0$ for all x because the graph of $y = e^x$ lies above the x-axis. So the turning points are $O(0,0)$ and $A(-2, 4e^{-2})$.

$$\frac{dy}{dx} = e^x(x^2 + 2x)$$

For the points of inflexion, you need to solve the equation

$$\frac{d^2y}{dx^2} = 0$$

Using the product rule again:

$$u = x^2 + 2x \qquad v = e^x$$

$$\frac{du}{dx} = 2x + 2 \qquad \frac{dv}{dx} = e^x$$

$$\frac{d^2y}{dx^2} = v\frac{du}{dx} + u\frac{dv}{dx} = (2x + 2)e^x + (x^2 + 2x)e^x$$

$$= (x^2 + 4x + 2)e^x$$

$e^x > 0 \Rightarrow$ possible points of inflexion occur at

$$x^2 + 4x + 2 = 0$$

$$x = \frac{-4 \pm \sqrt{(16 - 8)}}{2} = -2 \pm \sqrt{2}$$

So there may be points of inflexion at the two points

$$x = -2 - \sqrt{2}, \; y \approx 0.38$$

$$x = -2 + \sqrt{2}, \; y \approx 0.19$$

$$\frac{d^3y}{dx^3} = \frac{d}{dx}[(x^2 + 4x + 2)e^x]$$

$$= e^x\frac{d}{dx}(x^2 + 4x + 2) + (x^2 + 4x + 2)\frac{d}{dx}(e^x)$$

(by the product rule)

So: $\qquad \dfrac{d^3y}{dx^3} = e^x(2x + 4) + (x^2 + 4x + 2)e^x$

$$= e^x(x^2 + 6x + 6)$$

At $x = -2 + \sqrt{2}$ and at $x = -2 - \sqrt{2}$ you can show that $\dfrac{d^3y}{dx^3} \neq 0$

and therefore the points $(-2 - \sqrt{2}, 0.38)$ and $(-2 + \sqrt{2}, 0.19)$ are points of inflexion on the curve with equation $y = x^2e^x$.

Example 14
Find the equations of the tangent and normal to the curve with equation $y = x(x - 1)^{\frac{1}{2}}$ at the point $(5, 10)$.

First you need to find $\dfrac{dy}{dx}$ using the product rule:

$$u = x \qquad v = (x - 1)^{\frac{1}{2}}$$

$$\frac{du}{dx} = 1 \qquad \frac{dv}{dx} = \tfrac{1}{2}(x - 1)^{-\frac{1}{2}}$$

So:
$$\frac{dy}{dx} = v\frac{du}{dx} + u\frac{dv}{dx}$$

$$= (x-1)^{\frac{1}{2}} + \tfrac{1}{2}x(x-1)^{-\frac{1}{2}}$$

At $(5, 10)$:
$$\frac{dy}{dx} = 2 + \tfrac{5}{2} \times \tfrac{1}{2} = \tfrac{13}{4}$$

The equation of the tangent at $(5, 10)$ to $y = x(x-1)^{\frac{1}{2}}$ is
$$y - 10 = \tfrac{13}{4}(x-5)$$

that is,
$$13x - 4y - 25 = 0$$

The normal has gradient $-\tfrac{4}{13}$. The equation of the normal is
$$y - 10 = -\tfrac{4}{13}(x-5)$$

that is,
$$4x + 13y - 150 = 0$$

Exercise 2C

1 The sides of a square are increasing at a constant rate of $0.1\,\text{cm}\,\text{s}^{-1}$. Find the rate at which the area of the square is increasing when each side of the square is of length $10\,\text{cm}$.

2 A circular oil-slick is increasing in area at a constant rate of $3\,\text{m}^2\,\text{s}^{-1}$. Find the rate of increase of the radius of the slick at the instant when the area is $1200\,\text{m}^2$.

3 You are told that $A = 14x^2$ and that x is increasing at $0.5\,\text{cm}\,\text{s}^{-1}$. Find the rate of change of A at the instant when $x = 8\,\text{cm}$.

4 Each edge of a contracting cube is decreasing at a rate of $0.06\,\text{cm}\,\text{s}^{-1}$. Find the rate of decrease of (a) the volume (b) the outer surface area of the cube when the edges of the cube are each $4\,\text{cm}$.

5 The volume of an expanding sphere is increasing at a rate of $24\,\text{cm}^3\,\text{s}^{-1}$. Find the rate of increase of (a) the radius (b) the surface area when the radius of the sphere is $20\,\text{cm}$.

6 Given that $y = (3t-1)^2$ and $t = 4x^{\frac{1}{4}}$, find the value of $\dfrac{dy}{dx}$ when $x = 81$.

7 (a) Given that
$$y = \frac{\ln x}{x}, \quad x > 0$$
find the value of x for which $\dfrac{dy}{dx} = 0$.

(b)

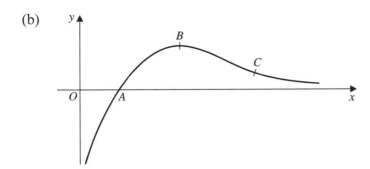

The diagram shows the curve with equation $y = \dfrac{\ln x}{x}$. The

curve crosses the x-axis at A; at B, $\dfrac{dy}{dx} = 0$; at C, $\dfrac{d^2y}{dx^2} = 0$.

Determine the coordinates of A, B and C.

8 Show that the curve with equation $y = \dfrac{e^{-x}}{x}$ has a maximum
value of $-c$ at $x = -1$.

9 Given that

$$y = \frac{2 - 8x}{x^2 + 2x}$$

find the values of x and of y when $\dfrac{dy}{dx} = 0$. Investigate the
nature of these stationary values of y.

10 When the depth of liquid in a container is x cm, the volume of
liquid is $x(x^2 + 25)$ cm^3. Liquid is added to the container at a
constant rate of $2\,\text{cm}^3\,\text{s}^{-1}$. Find the rate of change of the depth
of liquid at the instant when $x = 11$.

11 Express $\dfrac{2x}{x^2 - 4}$ in partial fractions. Given that $f(x) \equiv \dfrac{2x}{x^2 - 4}$,

find $f'(4)$ and $f''(0)$.

12 Given that

$$y = \frac{1}{x^2 + 4}$$

find the value of $\dfrac{d^2y}{dx^2}$ at $x = 1$.

13 Given that

$$y = \frac{e^x}{x + 1}$$

find, in terms of e, the values of $\dfrac{dy}{dx}$ and $\dfrac{d^2y}{dx^2}$ at $x = 1$.

14 If $y = xe^{-x}$, prove that

$$\frac{d^2y}{dx^2} + 2\frac{dy}{dx} + y = 0 \qquad\qquad \text{[E]}$$

15 If $y = x^n e^{ax}$, where n and a are constants, prove that

$$\frac{dy}{dx} - ay = \frac{ny}{x} \qquad\qquad \text{[E]}$$

16 Given that $y = 2e^{-4x} - e^{3x}$, prove that

(a) $y < 1$ for $x > 0$

(b) $\dfrac{d^2y}{dx^2} + 2\dfrac{dy}{dx} - 8y = -7e^{3x}$ $\qquad\qquad$ [E]

17 Find the equations of the tangent and the normal at the point $(3, 4)$ on the curve with equation

$$y = (x^2 + 7)^{\frac{1}{2}}$$

18

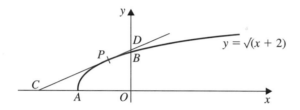

The curve with equation $y = \sqrt{(x + 2)}$ meets the x-axis at A and the y-axis at B, as shown.

(a) Find the coordinates of A and B.

At the point P on the curve, the x-coordinate is -1. The tangent to the curve at P meets the x-axis at C and the y-axis at D.

(b) Find an equation of the tangent.

(c) Determine the length of CD.

The normal at P to the curve meets the y-axis at E.

(d) Determine the distance BE.

19 Show that the line with equation $x + y = 0$ is the normal to the curve with equation $y = xe^x$ at the origin O.

20 Find the equations of the tangent and normal to the curve with equation $y = x \ln x$ at the point P, whose x-coordinate is e. The tangent and normal meet the x-axis at Q and R respectively. Find the length of QR and the area of $\triangle PQR$.

21 Given that $y = 4e^{3x} - 5e^{-3x}$, prove that

$$\frac{d^2y}{dx^2} = 9y$$

22 On the curve with equation $y = (3x + 1)^{\frac{1}{2}}$, the points P and Q have x-coordinates of 1 and 8 respectively. Find equations of the normals to the curve at P and Q.

23 Prove that the curve with equation $y = 4x^5 + \lambda x^3$ has two turning points for $\lambda < 0$ and none for $\lambda \geqslant 0$. Given that $\lambda = -\frac{5}{3}$, find the coordinates of the turning points and distinguish between them.

24 The curve with equation $y = e^x(px^2 + qx + r)$ is such that the tangents at $x = 1$ and $x = 3$ are parallel to the x-axis. The point $(0,9)$ is on the curve. Find the values of p, q and r.

25 (a) Find equations of the tangent and the normal at $(\frac{1}{2}, 1)$

on the curve with equation $y = \dfrac{x}{1 - x}$.

(b) Find the coordinates of the point where the normal meets the curve again.

26 For the curve with equation $y = \dfrac{x}{(x + 3)^2}$ find:

(a) the coordinates of the stationary point

(b) the coordinates of the point of inflexion.

27 For the curve with equation $y = \dfrac{2x}{1 + x^2}$,

(a) prove that
$$\frac{dy}{dx} = \frac{2(1 - x^2)}{(1 + x^2)^2}$$

(b) find the coordinates of the stationary points and distinguish between them.

28 The curve with equation $y = x^2 \ln x$ is defined for positive values of x.

(a) Determine the coordinates of the stationary point.

(b) Find an equation of the tangent at the point (e, e^2).

29 Given that
$$y = \frac{2x^2 - x + 1}{(x + 1)(x^2 + 1)}$$

express y in terms of partial fractions.

Hence determine the value of $\dfrac{dy}{dx}$ at $x = 1$.

30 Find the values of A, B and C for which
$$f(x) \equiv \frac{3x^2 - 2x + 1}{x^2(1 - x)} \equiv \frac{A}{1 - x} + \frac{B}{x^2} + \frac{C}{x}$$

Hence find $f'(2)$ and $f''(2)$.

2.7 Differentiating trigonometric functions

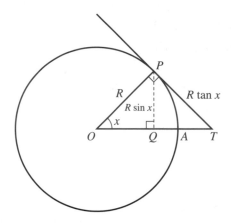

This circle, centre O and radius R, has the angle $POA = x$ radians, where $x < \dfrac{\pi}{2}$. The tangent to the circle at P meets OA produced at the point T. The radius OP is perpendicular to the tangent PT. The line PQ is perpendicular to OA. In $\triangle POQ$, $PQ = R \sin x$; in $\triangle OPT$, $PT = R \tan x$. Think about the areas of parts of the diagram. You can see that

$$\text{area } \triangle OAP < \text{area sector } OAP < \text{area } \triangle OPT$$

$$\tfrac{1}{2} R^2 \sin x < \qquad \tfrac{1}{2} R^2 x \qquad < \tfrac{1}{2} R^2 \tan x$$

So: $\qquad\qquad \sin x < \qquad\quad x \qquad\quad < \dfrac{\sin x}{\cos x}$

Divide by $\sin x$, which is positive, as $x < \dfrac{\pi}{2}$:

$$1 < \frac{x}{\sin x} < \frac{1}{\cos x}$$

As $x \to 0$, $\cos x \to 1$ and so $\dfrac{x}{\sin x} \to 1$
or you could say

$$\lim_{x \to 0} \left(\frac{x}{\sin x} \right) = 1$$

This limit is of vital importance in finding the derivative of $\sin x$ because if you consider

$$y = \sin x$$
$$y + \delta y = \sin(x + \delta x)$$
$$= \sin x \cos \delta x + \cos x \sin \delta x$$

[using the identity for $\sin(A + B)$]

When δx is small enough, $\cos \delta x \approx 1$ and since

$$\lim_{\delta x \to 0} \left(\frac{\delta x}{\sin \delta x} \right) = 1$$

$\sin \delta x \approx \delta x$.

So when δx is sufficiently small,

$$y + \delta y \approx \sin x + \delta x \cos x$$

$$\delta y \approx \sin x + \delta x \cos x - y$$

But $y = \sin x$

So: $\qquad \delta y \approx \delta x \cos x$ and $\dfrac{\delta y}{\delta x} \approx \cos x$

So: $\qquad \lim_{\delta x \to 0} \left(\dfrac{\delta y}{\delta x} \right) = \dfrac{dy}{dx} = \cos x$

■ $\dfrac{d}{dx}(\sin x) = \cos x$

Now $\cos x \equiv \sin\left(\dfrac{\pi}{2} - x\right)$. Write

$$y = \sin(\tfrac{\pi}{2} - x)$$

Put $t = \dfrac{\pi}{2} - x$; then $y = \sin t$ and differentiating gives

$$\frac{dt}{dx} = -1 \quad \text{and} \quad \frac{dy}{dt} = \cos t$$

By the chain rule.

$$\frac{dy}{dx} = \frac{dy}{dt} \cdot \frac{dt}{dx} = -\cos t = -\cos\left(\frac{\pi}{2} - x\right)$$

But $\cos\left(\dfrac{\pi}{2} - x\right) \equiv \sin x$ and so $\dfrac{dy}{dx} = -\sin x$

■ $\dfrac{d}{dx}(\cos x) = -\sin x$

In your examination, you need to know the derivatives of $\sin x$ and $\cos x$ but you will not be expected to prove the results just established. The derivation of these results has assumed that **the angle x is always measured in radians**. In all future work connected with differentiation and integration of trigonometric functions you must always take x to be measured in radians. Expressions such as $\sin x$, $\cos^2 \theta$, $\tan 3y$ all imply that x, θ and y are in radians and you should *never* use degrees in this level of work, except in practical trigonometry and, perhaps, in solving trigonometric equations, when you are instructed to give the answer in degrees.

You should memorise the following important results obtained using the chain rule:

- $\dfrac{d}{dx}(\sin nx) = n\cos nx$

- $\dfrac{d}{dx}(\cos nx) = -n\sin nx$

- $\dfrac{d}{dx}(\sin^n x) = n\sin^{n-1} x \cos x$

- $\dfrac{d}{dx}(\cos^n x) = -n\cos^{n-1} x \sin x$

The derivative of $\tan x$ is found using the quotient rule, because

$$y = \tan x = \frac{\sin x}{\cos x}$$

Take $\qquad\qquad u = \sin x \qquad v = \cos x$

Then: $\qquad\qquad \dfrac{du}{dx} = \cos x \qquad \dfrac{dv}{dx} = -\sin x$

$$\frac{dy}{dx} = \frac{v\dfrac{du}{dx} - u\dfrac{dv}{dx}}{v^2}$$

$$= \frac{\cos^2 x - (-\sin^2 x)}{\cos^2 x} = \frac{\cos^2 x + \sin^2 x}{\cos^2 x}$$

But $\cos^2 x + \sin^2 x \equiv 1$ and so:

$$\frac{d}{dx}(\tan x) = \frac{1}{\cos^2 x} = \sec^2 x$$

Similarly you can show that

$$\frac{d}{dx}(\cot x) = -\operatorname{cosec}^2 x$$

The derivative of $\sec x$ is found by using the chain rule:

$$y = \sec x = \frac{1}{\cos x} = (\cos x)^{-1}$$

Put $t = \cos x$ and then $y = t^{-1}$

$$\frac{dt}{dx} = -\sin x$$

$$\frac{dy}{dt} = -t^{-2} = \frac{-1}{t^2} = \frac{-1}{\cos^2 x}$$

So:
$$\frac{dy}{dx} = \frac{dy}{dt} \cdot \frac{dt}{dx} = \frac{\sin x}{\cos^2 x}$$

$$= \frac{1}{\cos x} \cdot \frac{\sin x}{\cos x}$$

$$= \sec x \tan x$$

Similarly you can show that

$$\frac{d}{dx}(\text{cosec}\, x) = -\text{cosec}\, x \cot x$$

You should learn all these results:

■ $\dfrac{d}{dx}(\sec x) = \sec x \tan x$

■ $\dfrac{d}{dx}(\text{cosec}\, x) = -\text{cosec}\, x \cot x$

■ $\dfrac{d}{dx}(\tan x) = \sec^2 x$

■ $\dfrac{d}{dx}(\cot x) = -\text{cosec}^2 x$

Example 15

Differentiate with respect to x:

(a) $\cos^4 x$ (b) $\tan(2x - 3)$ (c) $x^2 \sin 3x$

(a) Write $y = \cos^4 x$ and $\cos x = t$

Then:
$$y = t^4 \quad \text{and} \quad t = \cos x$$

$$\frac{dy}{dt} = 4t^3 \text{ and } \frac{dt}{dx} = -\sin x$$

Using the chain rule:
$$\frac{dy}{dx} = \frac{dy}{dt} \cdot \frac{dt}{dx} = -4t^3 \sin x$$

So:
$$\frac{dy}{dx} = -4\cos^3 x \sin x$$

(b) Write $y = \tan(2x - 3)$ and $2x - 3 = t$

Then:
$$y = \tan t \quad \text{and} \quad t = 2x - 3$$

$$\frac{dy}{dt} = \sec^2 t \text{ and } \frac{dt}{dx} = 2$$

Using the chain rule:
$$\frac{dy}{dx} = \frac{dy}{dt} \cdot \frac{dt}{dx} = \sec^2 t \times 2$$

So: $$\frac{dy}{dx} = 2 \sec^2(2x - 3)$$

(c) Take $y = x^2 \sin 3x$ and write

$$u = x^2, \qquad v = \sin 3x$$

Then: $$\frac{du}{dx} = 2x, \quad \frac{dv}{dx} = 3 \cos 3x$$

Using the product rule:

$$\frac{dy}{dx} = v\frac{du}{dx} + u\frac{dv}{dx}$$

So: $$\frac{dy}{dx} = 2x \sin 3x + 3x^2 \cos 3x$$

Example 16

Given that $y = \sin x + \cos 2x$, $0 \leqslant x \leqslant \pi$, find (a) the stationary values of y (b) the nature of these stationary values.

(a) By differentiation: $\quad \dfrac{dy}{dx} = \cos x - 2 \sin 2x$

But $$\sin 2x \equiv 2 \sin x \cos x$$

So: $\quad \dfrac{dy}{dx} = \cos x - 4 \sin x \cos x = \cos x(1 - 4\sin x)$

At stationary values of y, $\dfrac{dy}{dx} = 0$

So: $$\cos x = 0 \quad \text{or} \quad \sin x = \tfrac{1}{4}$$

$$x = \frac{\pi}{2} \quad \text{or} \quad 0.253 \quad \text{or} \quad 2.889$$

Stationary values are

$$y = 0 \quad \text{or} \quad 1.125 \quad \text{or} \quad 1.125$$

The coordinates of the stationary points are

$$\left(\frac{\pi}{2}, 0\right), \quad (0.253, 1.125), \quad (2.889, 1.125)$$

(b) By differentiation of $\dfrac{dy}{dx} = \cos x - 2 \sin 2x$, you have

$$\frac{d^2y}{dx^2} = -\sin x - 4 \cos 2x$$

At $x = 0.253$, $\quad \dfrac{d^2y}{dx^2} = -3.75 < 0$

At $x = \dfrac{\pi}{2}$, $\dfrac{d^2y}{dx^2} = 3 > 0$

At $x = 2.889$, $\dfrac{\mathrm{d}^2 y}{\mathrm{d}x^2} = -3.75 < 0$

At $(0.253, 1.125)$, $\dfrac{\mathrm{d}y}{\mathrm{d}x} = 0$, $\dfrac{\mathrm{d}^2 y}{\mathrm{d}x^2} < 0$

This stationary point is a *maximum*.

At $\left(\dfrac{\pi}{2}, 0\right)$, $\dfrac{\mathrm{d}y}{\mathrm{d}x} = 0$, $\dfrac{\mathrm{d}^2 y}{\mathrm{d}x^2} > 0$

This stationary point is a *minimum*.

At $(2.889, 1.125)$, $\dfrac{\mathrm{d}y}{\mathrm{d}x} = 0$, $\dfrac{\mathrm{d}^2 y}{\mathrm{d}x^2} < 0$

This stationary point is a *maximum*.

Exercise 2D

Differentiate with respect to x:

1 $\sin 3x$ **2** $\sin \frac{1}{2} x$ **3** $\cos 4x$ **4** $\cos \frac{2}{3} x$

5 $\tan 2x$ **6** $\tan \dfrac{x}{4}$ **7** $\sec 5x$ **8** $\operatorname{cosec} \frac{1}{2} x$

9 $\cot 6x$ **10** $\sec \dfrac{x}{2}$ **11** $\cot \dfrac{3x}{2}$ **12** $\operatorname{cosec} \dfrac{2x}{3}$

Find $\dfrac{\mathrm{d}y}{\mathrm{d}x}$ in each of these questions:

13 $y = \sin^2 x$ **14** $y = \sin^3 x$ **15** $y = \sqrt{(\sin x)}$

16 $y = \cos^4 x$ **17** $y = \cos^5 x$ **18** $y = (\cos x)^{\frac{1}{3}}$

19 $y = \tan^2 x$ **20** $y = \sqrt{(\tan x)}$ **21** $y = \dfrac{1}{\tan x}$

22 $y = \dfrac{1}{\sin^2 x}$ **23** $y = \dfrac{4}{\cos^4 x}$ **24** $y = \operatorname{cosec}^2 x$

Differentiate with respect to x:

25 $\sin^2 2x$ **26** $\cos^2 3x$ **27** $\tan^3 2x$

28 $\sec^2 2x$ **29** $\sin(3x + 5)$ **30** $\cos^2(2x - 4)$

31 $\tan^2(1 - 2x)$ **32** $\cot^2 3x$ **33** $\operatorname{cosec}^2 4x$

34 $\sin^2 x + \cos^2 x$ **35** $(\cot x - \operatorname{cosec} x)^2$ **36** $(\sin x - \cos x)^2$

Differentiate with respect to x:

37 $\sin^2 x \cos x$ **38** $\sin x \cos^3 x$ **39** $\tan x \sec x$

40 $\sec x \operatorname{cosec} x$ **41** $x \sin^2 x$ **42** $x^2 \cos x$

43 $x^2 \tan 2x$ **44** $\dfrac{x}{\sin x}$ **45** $\dfrac{\sin x}{x}$

46 $\dfrac{x^2}{\cos x}$ **47** $\dfrac{\cos 3x}{x^3}$ **48** $\tan 2x \sec 3x$

49 $\sin^2 x \cos^3 x$ **50** $e^x \sin x$ **51** $\dfrac{e^{2x}}{\cos^2 x}$

52 $\sin^2 x \ln x$ **53** $\dfrac{e^{2x}}{\sin x + \cos x}$ **54** $\dfrac{\ln(2x-5)}{\cos^2 3x}$

55 Given that $y = \sin 3x$, show that $\dfrac{d^2 y}{dx^2} = -9y$.

56 Given that $y = A\cos 2x + B\sin 2x$, where A and B are constants, show that

$$\frac{d^2 y}{dx^2} + 4y = 0$$

57 Find the value of $\sin\theta$ when $7\sec\theta - 3\tan\theta$ is a minimum.

58 Find equations of the tangent and normal to the curve with equation

$y = \sin x$, $0 < x < \dfrac{\pi}{2}$, at the point where $\sin x = \frac{3}{5}$.

59 Find the value of $\dfrac{d}{dx}[\sec^2 x + \tan^3 x]$ at $x = \dfrac{3\pi}{4}$.

60 Differentiate $\dfrac{1 + \cos 2x}{1 - \cos 2x}$.

61 Show that $\dfrac{d}{dx}\left[\dfrac{1 + \cot x}{1 - \cot x}\right] = \dfrac{2}{\sin 2x - 1}$.

62 Given that $y = (A + x)\cos x$, where A is a constant, show that $\dfrac{d^2 y}{dx^2} + y$ is independent of A.

63 The tangent to the curve with equation $y = \tan 2x$ at the point where $x = \dfrac{\pi}{8}$ meets the y-axis at the point Y. Find the distance OY, where O is the origin.

64 The normal to the curve with equation $y = \sec^2 x$ at the point $P\left(\dfrac{\pi}{4}, 2\right)$ meets the line $y = x$ at the point Q. Find PQ^2.

65 Find an equation of the normal to the curve with equation $y = e^x(\cos x + \sin x)$ at the point $(0, 1)$.

66 The volume of a cone with base radius $a\cos\theta$ and height $a\sin\theta$ is V, where

$$V = \tfrac{1}{3}\pi a^3 \cos^2\theta \sin\theta$$

Given that θ can vary and a is a constant, find the maximum value of V in terms of a.

67 Find the coordinates of the turning points of the curve with equation $y = e^{2x} \cos x$ in the interval $0 \leqslant x \leqslant 2\pi$ and distinguish between them.

68 Show that the function f given by

$$f(x) = \sin x - x \cos x$$

is increasing throughout the interval $0 \leqslant x \leqslant \dfrac{\pi}{2}$.

2.8 Differentiating relations given implicitly

Up to this point you have learned how to find $\dfrac{dy}{dx}$ from a function given **explicitly** as $y = f(x)$.

For example: $\qquad y = x^3 \qquad \dfrac{dy}{dx} = 3x^2$

But if $y = x^3$, then $x = y^{\frac{1}{3}}$ so that differentiating with respect to y gives

$$\frac{dx}{dy} = \tfrac{1}{3} y^{-\frac{2}{3}}$$

That is, $\qquad \dfrac{dx}{dy} = \dfrac{1}{3y^{\frac{2}{3}}} = \dfrac{1}{3x^2}$

since $x^2 = y^{\frac{2}{3}}$.

In this case then,

$$\frac{dy}{dx} = \frac{1}{\dfrac{dx}{dy}}$$

We can generalise this result by considering a small change δx in x giving rise to a small change δy in y.

Notice that $\dfrac{\delta y}{\delta x} \cdot \dfrac{\delta x}{\delta y} = 1$

So: $\qquad \dfrac{\delta y}{\delta x} = \dfrac{1}{\dfrac{\delta x}{\delta y}}$

You also know that

$$\lim_{\delta x \to 0} \left(\frac{\delta y}{\delta x} \right) = \frac{dy}{dx} \quad \text{and} \quad \lim_{\delta y \to 0} \left(\frac{\delta x}{\delta y} \right) = \frac{dx}{dy}$$

So it is generally true that:

■ $\dfrac{dy}{dx} = \dfrac{1}{\dfrac{dx}{dy}}$

Often you are not given y as a function of x. Instead, relations between the two variables x and y are given **implicitly**. For example:

$$x^2 + y^2 = 16x$$

$$\sin(x + y) = \cos y$$

These relations can be differentiated directly by using the chain rule and the product or quotient rules if required.

If, for example, you want to differentiate y^3 with respect to x, you can use the chain rule, first differentiating y^3 with respect to y to get $3y^2$, then differentiating y with respect to x, to get $\dfrac{dy}{dx}$.

Thus: $\qquad \dfrac{d}{dx}(y^3) = 3y^2 \dfrac{dy}{dx} \quad$ by the chain rule

Similarly: $\qquad \dfrac{d}{dx}(y^n) = ny^{n-1} \dfrac{dy}{dx} \quad$ by the chain rule

If you want to differentiate xy with respect to x, you can use the product rule and write

$$\dfrac{d}{dx}(xy) = y\dfrac{d}{dx}(x) + x\dfrac{d}{dx}(y)$$

$$= y.1 + x\dfrac{dy}{dx}$$

$$= y + x\dfrac{dy}{dx}$$

Example 17

Find $\dfrac{dy}{dx}$ in terms of x and y for (a) $x^2 + y^2 = 16x$,

(b) $\sin(x + y) = \cos y$.

(a) Differentiating with respect to x:

$$2x + 2y\dfrac{dy}{dx} = 16$$

So: $\qquad\qquad\qquad y\dfrac{dy}{dx} = 8 - x$

$$\dfrac{dy}{dx} = \dfrac{8 - x}{y}$$

(b) Using the chain rule:

$$\frac{d}{dx}[\sin(x+y)] = \cos(x+y)\left[\frac{d}{dx}(x+y)\right]$$

$$= \cos(x+y)\left[1+\frac{dy}{dx}\right]$$

$$\frac{d}{dx}(\cos y) = -\sin y\left(\frac{dy}{dx}\right)$$

So when you differentiate the relation $\sin(x+y) = \cos y$ with respect to x you get:

$$\left(1+\frac{dy}{dx}\right)\cos(x+y) = -\sin y\frac{dy}{dx}$$

That is: $\quad\quad \frac{dy}{dx}[\cos(x+y)+\sin y] = -\cos(x+y)$

$$\frac{dy}{dx} = \frac{-\cos(x+y)}{\cos(x+y)+\sin y}$$

Example 18

Find an equation of the normal at the point $(2,1)$ on the curve with equation $y^2 + 3xy = 2x^2$ 1.

First differentiate with respect to x to obtain:

$$2y\frac{dy}{dx} + 3\left(y + x\frac{dy}{dx}\right) = 4x$$

Taking $x = 2$, $y = 1$ gives

$$2\frac{dy}{dx} + 3\left(1 + 2\frac{dy}{dx}\right) = 8$$

So: $\quad\quad 2\frac{dy}{dx} + 3 + 6\frac{dy}{dx} = 8 \Rightarrow \frac{dy}{dx} = \frac{5}{8}$

So the gradient of the tangent to the curve at $(2,1)$ is $\frac{5}{8}$.

;Gradient of normal at $(2,1)$ is $-\frac{8}{5}$.

An equation of the normal is

$$y - 1 = -\frac{8}{5}(x - 2)$$

That is, $\quad\quad 5y + 8x = 21$

Example 19

Given that $y = a^x$, $a > 0$, find $\frac{dy}{dx}$.

Taking logs to the base e gives

$$\ln y = \ln a^x = x \ln a$$

Differentiating with respect to x, using the chain rule,

$$\frac{1}{y}\frac{dy}{dx} = \ln a$$

$\Rightarrow$
$$\frac{dy}{dx} = y \ln a = a^x \ln a$$

- **This is an important result, which you should memorise:**

$$\frac{d}{dx}(a^x) = a^x \ln a$$

2.9 Differentiating functions given parametrically

You will often come across relationships between the variables x and y where x and y are given in terms of another variable t, where t is called a **parameter**. For example, the curve with equation $4y = 3x^2$, which is a parabola, can also be represented by the equations

$$x = 2t, \qquad y = 3t^2$$

where t is a parameter taking all real values.

You can move from the **parametric equations** to the (x, y) equation (the cartesian equation), by eliminating t (if this is possible and simple). In this example, you know that

$$t = \frac{x}{2}$$

and
$$t^2 = \frac{y}{3}$$

So:
$$\frac{y}{3} = \left(\frac{x}{2}\right)^2$$

giving $4y = 3x^2$ as the cartesian equation.

Parametric equations are often used to reduce the amount of working needed to solve problems, particularly in coordinate geometry.

By using the chain rule, expressions for $\frac{dy}{dx}$ can be easily obtained:

$$\frac{dy}{dx} = \frac{dy}{dt} \cdot \frac{dt}{dx} = \frac{\frac{dy}{dt}}{\frac{dx}{dt}}$$

because $\frac{dt}{dx} = \frac{1}{\frac{dx}{dt}}$, as shown in section 2.8.

Example 20

Find the gradient at the point P where $t = -1$, on the curve given parametrically by

$$x = t^2 - t, \quad y = t^3 - t^2$$

Hence find equations of the tangent and normal at P.

Differentiating the x and y relations with respect to t,

$$\frac{\mathrm{d}x}{\mathrm{d}t} = 2t - 1 \qquad \frac{\mathrm{d}y}{\mathrm{d}t} = 3t^2 - 2t$$

$$\frac{\mathrm{d}y}{\mathrm{d}x} = \frac{\dfrac{\mathrm{d}y}{\mathrm{d}t}}{\dfrac{\mathrm{d}x}{\mathrm{d}t}} = \frac{3t^2 - 2t}{2t - 1}$$

At $t = -1$, $x = 2$, $y = -2$, and

$$\frac{\mathrm{d}y}{\mathrm{d}x} = \frac{5}{-3} = -\tfrac{5}{3}$$

An equation of the tangent at P is

$$y + 2 = -\tfrac{5}{3}(x - 2)$$

The normal has gradient $\tfrac{3}{5}$ and its equation is

$$y + 2 = \tfrac{3}{5}(x - 2)$$

Example 21

Find an equation of the tangent at the point P where $t = \dfrac{\pi}{4}$, on the curve with parametric equations

$$x = 2\cos t - \cos 2t, \quad y = 2\sin t - \sin 2t, \quad 0 \leqslant t < \pi$$

Differentiating both x and y with respect to t:

$$\frac{\mathrm{d}x}{\mathrm{d}t} = -2\sin t + 2\sin 2t, \quad \frac{\mathrm{d}y}{\mathrm{d}t} = 2\cos t - 2\cos 2t$$

Using the chain rule:

$$\frac{\mathrm{d}y}{\mathrm{d}x} = \frac{\dfrac{\mathrm{d}y}{\mathrm{d}t}}{\dfrac{\mathrm{d}x}{\mathrm{d}t}} = \frac{2\cos t - 2\cos 2t}{2\sin 2t - 2\sin t} = \frac{\cos t - \cos 2t}{\sin 2t - \sin t}$$

At $t = \dfrac{\pi}{4}$, $x = \sqrt{2}$, $y = \sqrt{2} - 1$, $\dfrac{\mathrm{d}y}{\mathrm{d}x} = \dfrac{\frac{1}{\sqrt{2}}}{1 - \frac{1}{\sqrt{2}}} = \dfrac{1}{\sqrt{2} - 1}$

$$\frac{\mathrm{d}y}{\mathrm{d}x} = \frac{1}{\sqrt{2} - 1} \times \frac{\sqrt{2} + 1}{\sqrt{2} + 1} = \frac{\sqrt{2} + 1}{2 - 1} = \sqrt{2} + 1$$

An equation of the tangent at P is:

$$y - (\sqrt{2} - 1) = (\sqrt{2} + 1)(x - \sqrt{2})$$
$$y - \sqrt{2} + 1 = x(\sqrt{2} + 1) - 2 - \sqrt{2}$$

That is,
$$x(\sqrt{2} + 1) - y = 3$$

Exercise 2E

Find $\dfrac{dy}{dx}$ for each of these relations:

1 $y^2 = 2x + 1$ **2** $x^2 + y^2 = 4$ **3** $xy^2 = 16$

4 $x^2 + 2xy + 3y^2 = 6$ **5** $\sin x \cos y = 1$ **6** $e^{xy} = 4$

7 $\sin(x + y) = \frac{1}{2}$ **8** $\ln y \ln x = 3$ **9** $e^x \ln y = x$

10 $y(x + y) = 12$

Find $\dfrac{dy}{dx}$ for each of the following where t, θ and u are parameters; a, b and c are constants.

11 $x = 3t^2$, $y = 2t$ **12** $x = t^2$, $y = t^3$

13 $x = ct$, $y = \dfrac{c}{t^2}$ **14** $x = a \cos t$, $y = b \sin t$

15 $x = a \sec t$, $y = a \tan t$ **16** $x = e^t \cos t$, $y = e^t \sin t$

17 $x = t \cos t$, $y = t \sin t$

18 $x = a(\theta - \cos \theta)$, $y = a(1 + \sin \theta)$

19 $x = e^u$, $y = e^u - e^{-u}$ **20** $x = \sin^2 \theta$, $y = \cos \theta \sin \theta$

21 Find equations of the tangents to the curve with equation $y = 2^x$ at the points P and Q where $x = 2$ and $x = 5$ respectively. These tangents meet at the point R. Find the x-coordinate of R.

22 Find an equation of the normal to the curve with parametric equations $x = 3t - t^3$, $y = t^2$ at the point where $t = 2$.

23 A curve is given by the equations
$$x = t^2, \quad y = t^5$$
where t is a parameter.
Find equations of the tangent and normal to the curve at the point P where $t = -1$.
Find a cartesian equation of the curve.

24 Find an equation of the tangent to the curve with equations
$$x = (1 - \tfrac{1}{3}t^3), \quad y = 4t^2$$
at the point P, where $t = -2$.

25 The tangents at the points P and Q in the first quadrant, where $x = 1$ and $x = 4$, respectively to the curve with equation $y^2 = 4x$ meet at the point R. Find the coordinates of R.

26 Find an equation of the tangent at $(2,1)$ to the curve with equation

$$y(x + y)^2 = 3(x^3 - 5)$$

27 Find $\dfrac{dy}{dx}$ in terms of t for the curve with parametric equations

$$x = \frac{t}{1 - t}, \quad y = \frac{t^2}{1 - t}$$

Deduce the equation of the normal at the point where $t = \frac{1}{2}$.

28 For the curve with equation $xy(x + y) = 84$, find $\dfrac{dy}{dx}$ at $(3,4)$.

29 Differentiate with respect to x:

(a) 10^x (b) 2^{x^2} (c) 5^{-x}

30 For the curve with equation $2\cos y \sin x = 1$, find an equation of the normal at $\left(\dfrac{\pi}{4}, \dfrac{\pi}{4}\right)$.

SUMMARY OF KEY POINTS

1 If y is a function of t and t is a function of x, then

$$\frac{dy}{dx} = \frac{dy}{dt} \cdot \frac{dt}{dx}$$

This is called the chain rule.

2 If $y = uv$, where u and v are functions of x, then

$$\frac{dy}{dx} = v\frac{du}{dx} + u\frac{dv}{dx}$$

This is called the product rule.

3 If $y = \dfrac{u}{v}$, where u and v are functions of x, then

$$\frac{dy}{dx} = \frac{v\dfrac{du}{dx} - u\dfrac{dv}{dx}}{v^2}$$

This is called the quotient rule.

4 The chain rule is used to find connected rates of change; for example:

$$\frac{dV}{dt} = \frac{dV}{dr} \cdot \frac{dr}{dt}$$

5 You should know that $\lim\limits_{x \to 0} \left(\dfrac{\sin x}{x} \right) = 1$.

6 Memorise these standard formulae for derivatives:
   (a, b, k, n are constants)

y or $f(x)$	$\dfrac{dy}{dx}$ or $f'(x)$
x^n	nx^{n-1}
$(ax + b)^n$	$na(ax + b)^{n-1}$
e^x	e^x
e^{ax+b}	ae^{ax+b}
a^x	$a^x(\ln a)$
$\ln x$	$\dfrac{1}{x}$
$\ln(ax + b)$	$\dfrac{a}{ax + b}$
$\sin x$	$\cos x$
$\cos x$	$-\sin x$

$\tan x$	$\sec^2 x$
$\cot x$	$-\mathrm{cosec}^2 x$
$\sec x$	$\sec x \tan x$
$\mathrm{cosec}\ x$	$-\mathrm{cosec}\ x \cot x$
$\sin kx$	$k \cos kx$
$\sin^n x$	$n \sin^{n-1} x \cos x$
$\sin^n kx$	$nk \sin^{n-1} kx \cos kx$
$u \pm v$	$\dfrac{\mathrm{d}u}{\mathrm{d}x} \pm \dfrac{\mathrm{d}v}{\mathrm{d}x}$
uv	$v\dfrac{\mathrm{d}u}{\mathrm{d}x} + u\dfrac{\mathrm{d}v}{\mathrm{d}x}$
$\dfrac{u}{v}$	$\dfrac{v\dfrac{\mathrm{d}u}{\mathrm{d}x} - u\dfrac{\mathrm{d}v}{\mathrm{d}x}}{v^2}$
y^n	$ny^{n-1}\dfrac{\mathrm{d}y}{\mathrm{d}x}$
xy	$y + x\dfrac{\mathrm{d}y}{\mathrm{d}x}$

7 $\dfrac{\mathrm{d}y}{\mathrm{d}x} = \dfrac{1}{\dfrac{\mathrm{d}x}{\mathrm{d}y}}$

8 If the equation of a curve is given in terms of its parametric equations, so that $x = \mathrm{f}(t)$ and $y = \mathrm{g}(t)$ then:

$$\frac{\mathrm{d}y}{\mathrm{d}x} = \frac{\dfrac{\mathrm{d}y}{\mathrm{d}t}}{\dfrac{\mathrm{d}x}{\mathrm{d}t}}$$

Coordinate geometry in the *xy*-plane

3

The circle is often defined as the path of a point which moves in a plane so that the distance, called the **radius**, from a fixed point, called the **centre**, is constant. This chapter deals with the geometry of the circle in some detail, then introduces some other curves, given in either **cartesian** or **parametric** form and shows how in some cases you can convert from one form to the other.

3.1 The circle

Consider first a circle whose centre is at the origin O and whose radius is a.

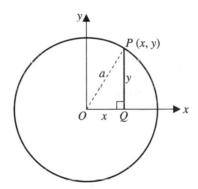

Take any point $P(x, y)$ on the circle and complete the right-angled triangle OPQ, as shown, where $OQ = x$ and $PQ = y$. By Pythagoras' theorem you have

$$OP^2 = OQ^2 + PQ^2$$

That is:
$$a^2 = x^2 + y^2$$

Clearly, all points on the circumference of the circle which, by definition, are at a fixed distance a from O satisfy this equation.

■ **The cartesian equation of the circle centre O and radius a is $x^2 + y^2 = a^2$.**

Consider now a circle, of radius r, whose centre C is at the point (α, β).

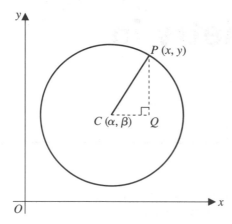

Take any point $P(x, y)$ on the circle and complete the right-angled triangle PCQ as shown, where CQ is parallel to the x-axis and PQ is parallel to the y-axis. Since for all positions of P on the circle $CP = r$, and, by Pythagoras' theorem, $CP^2 = CQ^2 + PQ^2$, you have

$$r^2 = (x - \alpha)^2 + (y - \beta)^2$$

because $CQ = x - \alpha$ and $PQ = y - \beta$.

- **The cartesian equation of the circle, centre (α, β) and radius r, is**

$$(x - \alpha)^2 + (y - \beta)^2 = r^2$$

When multiplied out, the equation of this circle is

$$x^2 + y^2 - 2\alpha x - 2\beta y + \alpha^2 + \beta^2 - r^2 = 0$$

- **The general equation of a circle is usually given in the form**

$$x^2 + y^2 + 2gx + 2fy + c = 0$$

where g, f and c are constants.

Notice that:
(i) the equation is second degree because x^2 and y^2 are the highest power terms
(ii) the coefficients of x^2 and y^2 are the same
(iii) there is no term in xy.

If you use the identities

$$x^2 + 2gx + g^2 \equiv (x + g)^2$$

$$y^2 + 2fy + f^2 \equiv (y + f)^2$$

you can write the equation $x^2 + y^2 + 2gx + 2fy + c = 0$ as

$$x^2 + 2gx + g^2 + y^2 + 2fy + f^2 = g^2 + f^2 - c$$

that is: $\qquad (x + g)^2 + (y + f)^2 = \left[\sqrt{(g^2 + f^2 - c)}\right]^2$

If you compare this with the equation of the circle

$$(x - \alpha)^2 + (y - \beta)^2 = r^2$$

whose radius is r and centre (α, β), you see that

■ **the circle with equation $x^2 + y^2 + 2gx + 2fy + c = 0$ has centre $(-g, -f)$ and radius $\sqrt{(g^2 + f^2 - c)}$.**

Example 1

Find, in cartesian form, an equation of the circle

(a) with centre the origin and radius 6
(b) with centre $(-3, 5)$ and radius 4.

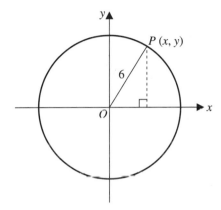

(a) For any point $P(x, y)$ on the circle,

$$OP^2 = x^2 + y^2 \quad \text{and} \quad OP = 6$$
$$\Rightarrow \quad x^2 + y^2 = 6^2$$

An equation is $x^2 + y^2 = 36$.

(b) For any point $P(x, y)$ on the circle, where $C(-3, 5)$ is the centre, you have

$$PC^2 = PQ^2 + QC^2$$

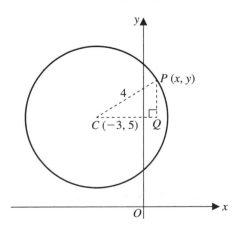

But $PC = 4$, the radius of the circle.

So: $$PQ = y - 5 \quad \text{and} \quad QC = x + 3$$

An equation of the circle is $(y - 5)^2 + (x + 3)^2 = 16$

or $$x^2 + y^2 + 6x - 10y + 18 = 0$$

Example 2

Find the coordinates of the centre and the radius of the circle with equation

$$2x^2 + 2y^2 - 8x + 12y + 1\tfrac{1}{2} = 0$$

Divide by 2 to obtain on rearranging:

$$x^2 - 4x + y^2 + 6y + \tfrac{3}{4} = 0$$

Complete the square for $x^2 - 4x$ and $y^2 + 6y$ by adding 4 and 9 respectively:

$$x^2 - 4x + 4 + y^2 + 6y + 9 + \tfrac{3}{4} = 4 + 9$$

That is: $$(x - 2)^2 + (y + 3)^2 = 12\tfrac{1}{4} = \left(\tfrac{7}{2}\right)^2$$

So an equation of the circle is $(x - 2)^2 + (y + 3)^2 = \left(\tfrac{7}{2}\right)^2$.

That is, the centre of circle is at $(2, -3)$ and the radius of the circle is $\tfrac{7}{2} = 3\tfrac{1}{2}$.

Example 3

Find, in cartesian form, an equation of the circle which has the points $A(2, 5)$ and $B(10, -1)$ as the ends of a diameter.

The mid-point C of AB is $\left[\dfrac{2 + 10}{2}, \dfrac{5 - 1}{2}\right] = (6, 2)$ and this is the centre of the required circle.

Also: $$AC^2 = (2 - 6)^2 + (5 - 2)^2 = 16 + 9 = 25$$

But AC is the radius of the required circle. So:

$$\text{radius of circle} = 5$$

The equation of the required circle is

$$(x - 6)^2 + (y - 2)^2 = 25$$

Example 4

Find, in cartesian form, an equation of the circle passing through the points $A(1, 6)$, $B(3, 2)$ and $C(2, 3)$.

Coordinate geometry in the *xy*-plane **65**

The centre of the required circle lies on the perpendicular bisector of AB.

The mid-point of AB is $\left(\dfrac{1+3}{2}, \dfrac{6+2}{2}\right) = (2, 4)$.

The gradient of AB is $\dfrac{6-2}{1-3} = -2$.

So the gradient of the perpendicular bisector is
$$\dfrac{-1}{-2} = \tfrac{1}{2}$$

and its equation is
$$y - 4 = \tfrac{1}{2}(x - 2) \qquad\qquad (1)$$

The centre of the required circle lies also on the perpendicular bisector of BC.

The mid-point of BC is $\left(\dfrac{3+2}{2}, \dfrac{2+3}{2}\right) = \left(2\tfrac{1}{2}, 2\tfrac{1}{2}\right)$.

The gradient of BC is $\dfrac{3-2}{2-3} = -1$.

So the gradient of the perpendicular bisector is
$$\dfrac{-1}{-1} = 1$$

and its equation is
$$y - 2\tfrac{1}{2} = 1\left(x - 2\tfrac{1}{2}\right) \quad\Rightarrow\quad y = x \qquad\qquad (2)$$

You now need to solve equations (1) and (2) simultaneously to find the coordinates of the centre of the circle.

Putting $y = x$ in (1) gives
$$x - 4 = \tfrac{1}{2}(x - 2)$$
$$x - 4 = \tfrac{1}{2}x - 1$$
$$\tfrac{1}{2}x = 3 \quad\Rightarrow\quad x = 6$$

The centre of the circle is the point $(6, 6)$.

The radius of the circle is the distance from $(6, 6)$ to A:
$$\sqrt{\left[(6 - 1)^2 + (6 - 6)^2\right]} = \sqrt{25} = 5$$

So an equation of the circle is
$$(x - 6)^2 + (y - 6)^2 = 25$$

Another method of solving the problem in example 4 is to take the required circle as

$$x^2 + y^2 + 2gx + 2fy + c = 0$$

and form three equations by substituting the coordinates of A, B and C in turn into the equation. You could then solve the equations for g, f and c.

The equations are

$$2g + 12f + c = -37 \tag{1}$$

$$6g + 4f + c = -13 \tag{2}$$

$$4g + 6f + c = -13 \tag{3}$$

From (2) and (3) by subtraction you get $f = g$.

From (1) and (2) by subtraction you get $4g - 8f = 24$.

Hence $f = g = -6$ and $c = 47$ by substitution, giving the equation of the circle as

$$x^2 + y^2 - 12x - 12y + 47 = 0$$

The equations and properties of tangents to a circle

Because a circle is symmetrical it has certain properties. You should know that for any circle:

(a) Two tangents can be drawn from any external point T to the circle and these tangents are equal in length.

(b) The radius drawn to the **point of contact** of a tangent with the circle is at right angles to the tangent.

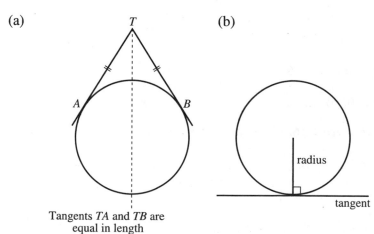

(a)

(b)

Tangents TA and TB are
equal in length

radius

tangent

Consider the circle with equation $x^2 + y^2 = a^2$ and a general point (h, k) on the circumference of the circle.

Differentiating the equation of the circle with respect to x gives

$$2x + 2y \frac{dy}{dx} = 0$$

That is:

$$\frac{dy}{dx} = -\frac{x}{y}$$

At (h, k):

$$\frac{dy}{dx} = -\frac{h}{k}$$

An equation of the tangent to the circle at (h, k) is

$$y - k = -\frac{h}{k}(x - h)$$

(see Book P1, page 144).

That is:

$$hx + ky = h^2 + k^2$$

But the point (h, k) lies on the circle, so

$$h^2 + k^2 = a^2$$

- **The tangent at (h, k) has equation $hx + ky = a^2$.**

By using a similar method you can show that the tangent to the circle $x^2 + y^2 + 2gx + 2fy + c = 0$ at the point (h, k) has equation

$$hx + ky + g(h + x) + f(k + y) + c = 0$$

Note: It will help you to check this equation when working examples if you appreciate that from the circle equation $x^2(= xx)$ is replaced by xh, $y^2(= yy)$ by yk, $2gx[= g(x + x)]$ by $g(x + h)$ and $2fy[= f(y + y)]$ by $f(y + k)$. Also, you will find later that this check works equally well when you are writing down the equations of tangents to a parabola, ellipse or hyperbola.

Example 5

The variable point $P(x, y)$ moves in such a way that $AP^2 = 4BP^2$ where A is the point $(1, 3)$ and B is the point $(4, -3)$.

(a) Show that P lies on the circle with equation $x^2 + y^2 - 10x + 10y + 30 = 0$.
(b) Find an equation of the tangent at $(1, -3)$ to the circle.
(c) The line OT is a tangent to the circle, where O is the origin and T lies on the circle. Calculate the length of OT.

(a) $AP^2 = (x - 1)^2 + (y - 3)^2$ and $BP^2 = (x - 4)^2 + (y + 3)^2$

Since $AP^2 = 4BP^2$ you have

$$x^2 - 2x + 1 + y^2 - 6y + 9 = 4x^2 - 32x + 64 + 4y^2 + 24y + 36$$
$$0 = 3x^2 + 3y^2 - 30x + 30y + 90$$

That is, $x^2 + y^2 - 10x + 10y + 30 = 0$ is the locus of P.

(b) Differentiating with respect to x gives

$$2x + 2y \frac{dy}{dx} - 10 + 10 \frac{dy}{dx} = 0$$

At the point $(1, -3)$ on the circle you have

$$2 - 6 \frac{dy}{dx} - 10 + 10 \frac{dy}{dx} = 0 \quad \Rightarrow \quad \frac{dy}{dx} = 2$$

An equation of the tangent to the circle at the point $(1, -3)$ is

$$y + 3 = 2(x - 1) \quad \Rightarrow \quad 2x - y = 5$$

(c) The circle equation $x^2 - 10x + y^2 + 10y + 30 = 0$ can be rearranged as

$$(x - 5)^2 + (y + 5)^2 = 20$$

which means that the radius is $\sqrt{20}$ and the centre C is at $(5, -5)$.

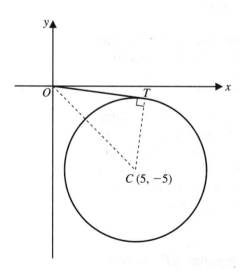

Referring to the diagram and using Pythagoras' theorem you have

$$OT^2 = OC^2 - CT^2$$
$$= 5^2 + 5^2 - 20$$

since CT^2 is the square of the radius of the circle .

Hence $OT^2 = 30$ and the length of the tangent OT is $\sqrt{30}$ units.

Exercise 3A

1 Write down the coordinates of the centre and the radius of the circle with equation
   (a) $x^2 + y^2 - 4x - 6y - 12 = 0$
   (b) $x^2 + y^2 + 6x + 4y - 36 = 0$
   (c) $2x^2 + 2y^2 - 2x - 2y - 1 = 0$
   (d) $x^2 + y^2 - 10x + 14y + 58 = 0$
   (e) $25x^2 + 25y^2 - 10x + 60y + 33 = 0$

2 Find an equation of the circle whose centre is at C and whose radius is r when
   (a) C is $(0, 0)$, $r = 6$ (b) C is $(0, -7)$, $r = 2$
   (c) C is $(3, -4)$, $r = 5$ (d) C is $\left(-\frac{1}{2}, -\frac{1}{3}\right)$, $r = 1$
   (e) C is $(-2, -3)$, $r = \sqrt{10}$

3 Given that AB is a diameter of a circle and the coordinates of A and B are as below, find an equation of the circle.
   (a) $A = (1, 0)$, $B = (3, 0)$ (b) $A = (8, 0)$, $B = (0, -6)$
   (c) $A = (5, -3)$, $B = (-3, 5)$

4 Prove that the circle which has $A(x_1, y_1)$ and $B(x_2, y_2)$ at the ends of a diameter AB has equation
   $$(x - x_1)(x - x_2) + (y - y_1)(y - y_2) = 0$$
   Now check the answers you got in question 3 by using the formula.

5 Find an equation of the circle passing through the points:
   (a) $(0, 0)$, $(0, 4)$, $(6, 0)$ (b) $(-2, 4)$, $(5, 5)$, $(6, 4)$
   (c) $(0, -2)$, $(-7, 5)$, $(-3, 7)$

6 A is the point $(3, 2)$ and B is the point $(2, 1)$. Find an equation for
   (a) the circle on AB as a diameter
   (b) the circle that passes through A, B and the origin O.

7 Find an equation of the tangent at $A(x_1, y_1)$ to the circle with equation $f(x, y) = 0$ where
   (a) $f(x, y) \equiv x^2 + y^2 - 25$, $x_1 = -3$, $y_1 = 4$
   (b) $f(x, y) \equiv x^2 + y^2 - 6x + 5y + 2$, $x_1 = 2$, $y_1 = 1$
   (c) $f(x, y) \equiv x^2 + y^2 + 8x - 6y + 8$, $x_1 = 0$, $y_1 = 2$

8 Find an equation of the normal at $A(x_1, y_1)$ to the circle with equation $f(x, y) = 0$, where the data are as given in question 7.

9 Find the least and the greatest distances of the origin from the circumference of the circle with equation
   $$(x - 12)^2 + (y - 5)^2 = 25$$

10 Show that the line with equation $2x - 3y + 26 = 0$ is a tangent to the circle with equation $x^2 + y^2 - 4x + 6y - 104 = 0$. If T is the point of contact of the tangent with the circle, find the equation of the normal to the circle at T. Find also the coordinates of the point on the circle which is diametrically opposite to T.

11 The line with equation $y = mx$ is a tangent to the circle with equation $x^2 + y^2 - 6x - 6y + 17 = 0$. Find the possible values of m.

12 Prove that the circle with equation
$x^2 + y^2 - 2ax - 2by + b^2 = 0$ touches the y-axis.
Hence, or otherwise, find equations of the two circles which pass through the points $(1, 2)$ and $(2, 3)$ and which touch the y-axis. Find the distance between their centres.

13 The point $P(x, y)$ moves so that its distance from the point $(3, 4)$ is twice its distance from the origin O. Find, in cartesian form, an equation for the path of P.

14 Show that the circle with equation $x^2 + y^2 - 2ax - 2ay + a^2 = 0$ touches both the x-axis and the y-axis. Hence show that there are two circles that pass through the point $(2, 4)$, just touching both the x- and y-axes.
Find an equation of the tangent to each circle at the point $(2, 4)$.

15 Given that the line with equation $y = x + c$ is a tangent to the circle with equation $x^2 + y^2 + 10x - 12y + 11 = 0$, find
(a) the possible values of c
(b) the coordinates of the corresponding points of contact of the tangent with the circle.

16 The points A and B have coordinates $(6, 4)$ and $(1, -8)$ respectively. Find the coordinates of the centre and the radius of the circle drawn on AB as diameter.

17

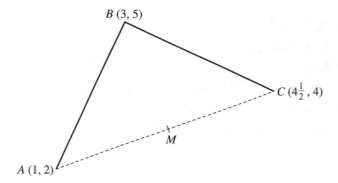

For the triangle ABC, show that

(a) AB is perpendicular to BC

(b) $BM = CM$, where M is the mid-point of AC.

Find also an equation of the circle passing through A, B and C.

18 The centre of a square is at $(3, 4)$ and one of its vertices is at $(7, 1)$. Find the coordinates of the other three vertices.

Find also an equation of the circle which passes through each vertex of the square.

3.2 Sketching curves given by cartesian equations

Often when you are reading questions and building solutions to them you will find you need to sketch the graphs of a variety of simple functions. There is no need to plot and draw a graph accurately unless you are specifically instructed to do so. You should, however, be able to sketch a curve whose equation is given. A review of what you need to do follows in this section.

The technique of curve sketching is used not only when demanded by an examination question. You should always use it in questions where you are asked to calculate the area under a curve. Unless you can see a picture of the curve first, and so know what area it is you are trying to calculate, it is very difficult to work out the correct limits for the integration and so to obtain a correct answer.

The form of some graphs should be familiar to you. The graph of $y = ax + b$ is a straight line. So a sketch of $y = 3x + 6$ looks like this:

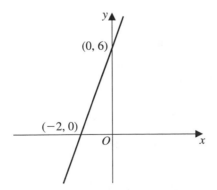

The line has positive gradient and meets the coordinate axes at the points $(-2, 0)$ and $(0, 6)$ as shown.

A sketch of $y = -4x - 8$ looks like this:

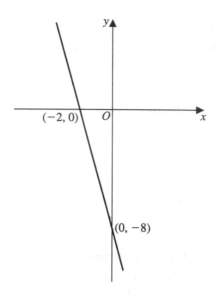

The line has negative gradient and meets the coordinate axes at $(-2, 0)$ and $(0, -8)$ as shown.

The graph of $y = ax^2 + bx + c$ is always a parabola.

When $a > 0$ the curve has a 'valley' shape: it is of the $\bigcup$ form.

When $a < 0$ the curve has a 'hill' shape: It is of the $\bigcap$ form.

Any curve of the form $y = f(x)$ cuts the *x*-axis where $y = 0$, so the graph of $y = ax^2 + bx + c$ cuts the *x*-axis where $ax^2 + bx + c = 0$. That is, where

$$x = \frac{-b \pm \sqrt{(b^2 - 4ac)}}{2a}$$

as shown in section 1.10 of Book P1.

If $b^2 - 4ac < 0$, then the equation $ax^2 + bx + c = 0$ has no real roots because a negative number has no real square root. That is, the graph of $y = ax^2 + bx + c$ does *not* cross the *x*-axis.

If $b^2 - 4ac = 0$, then $x = \frac{-b}{2a}$. That is, the equation $ax^2 + bx + c = 0$ has just *one* root. So the graph of $y = ax^2 + bx + c$ just touches the *x*-axis.

If $b^2 - 4ax > 0$, then equation $ax^2 + bx + c = 0$ has two real roots and so the graph of $y = ax^2 + bx + c$ cuts the *x*-axis at two distinct points.

The possibilities are summarised in the diagrams.

$a>0$ and $b^2-4ac<0$

$a>0$ and $b^2-4ac=0$

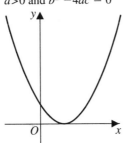

$a>0$ and $b^2-4ac>0$

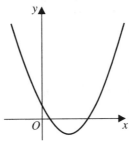

$a<0$ and $b^2-4ac<0$

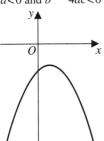

$a<0$ and $b^2-4ac=0$

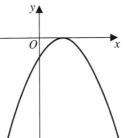

$a<0$ and $b^2-4ac>0$

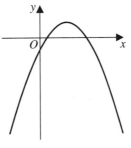

So a sketch of $y = x^2 - x - 2 = (x+1)(x-2)$ looks like this:

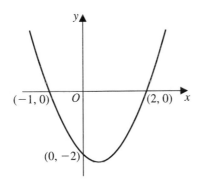

A sketch of $y = 6 + x - x^2 = (3-x)(2+x)$ looks like this:

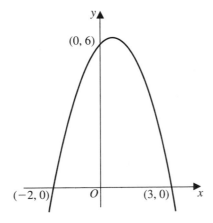

The graph of $y = ax^3 + bx^2 + cx + d$ either consists of one 'hill' and one 'valley', that is it has one maximum point and one minimum point, or sometimes these two merge to form a point of inflexion. In the equation $y = ax^3 + bx^2 + cx + d$, the ax^3 term dominates. So if $a > 0$, then:

as $x \to +\infty$, $y \to +\infty$

and as $x \to -\infty$, $y \to -\infty$

The graph therefore takes the form ⌒⌄ or ╱ .

If $a < 0$, then:

as $x \to +\infty$, $y \to -\infty$

and as $x \to -\infty$, $y \to +\infty$

The graph therefore takes the form ⌣⌒ or ╲ .

Here are some examples:

The graph $y = (x + 3)(x - 1)(x - 4)$

i.e. $\qquad y = x^3 - 2x^2 - 11x + 12$

looks like this:

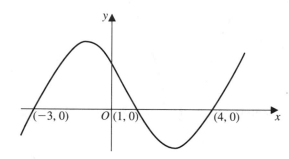

The graph of $y = (2 - x)(3 + x)(1 + x)$

i.e. $\qquad y = 6 + 5x - 2x^2 - x^3$

looks like this:

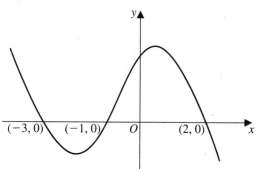

The graph of $y = (x - 1)^3$

i.e. $\qquad y = x^3 - 3x^2 + 3x - 1$

looks like this:

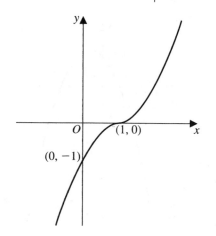

However, the shape of the graphs of most equations will be unfamiliar to you and so you need to have some general strategy for sketching a curve. There are a number of steps that you need to take.

(1) Where does the curve cut the axes? To find out where it cuts the *x*-axis you put $y = 0$ in the equation of the curve. To find out where it cuts the *y*-axis you put $x = 0$ in the equation of the curve.

(2) Are there any asymptotes and where do they occur?

(3) What happens as $x \to \pm\infty$?

(4) Has the graph any symmetry? You should know from Book P2 that $f(x)$ is an even function if $f(-x) = f(x)$. If $f(x)$ is an even function then the graph of $y = f(x)$ is symmetrical about the *y*-axis. Similarly, if the equation is $x = g(y)$ and $g(y)$ is an even function, then the graph of $x = g(y)$ is symmetrical about the *x*-axis. If $f(x)$ is an odd function, i.e. $f(-x) = -f(x)$, then the graph of $y = f(x)$ has rotational symmetry of $180°$ about the origin.

(5) Are there any values of *x* for which $f(x)$ is not defined?

(6) Where do the stationary points occur and are they maxima, minima or points of inflexion?

Example 6

Sketch the curve with equation $y = \dfrac{x + 2}{x - 2}$.

(1) When $x = 0$, $y = -1$

So the curve cuts the *y*-axis at $(0, -1)$.

When $y = 0$, $x + 2 = 0 \Rightarrow x = -2$

So the curve cuts the *x*-axis at $(-2, 0)$.

(2) $x = 2 \Rightarrow x - 2 = 0$

So at $x = 2$, *y* is undefined. Thus $x = 2$ is an asymptote parallel to the *y*-axis.

(3) $\dfrac{x + 2}{x - 2} = 1 + \dfrac{4}{x - 2}$

So as $x \to \pm\infty$, $\dfrac{4}{x - 2} \to 0$ and $y \to 1$

So $y = 1$ is an asymptote, parallel to the *x*-axis.

(4) $\dfrac{x + 2}{x - 2}$ is neither odd nor even.

(5) The curve exists for all *x* except $x = 2$.

(6) $\dfrac{\mathrm{d}y}{\mathrm{d}x} = \dfrac{(x-2)\cdot 1 - (x+2)\cdot 1}{(x-2)^2}$

$\qquad = -\dfrac{4}{(x-2)^2} < 0$ for all x except $x = 2$, where it is undefined.

So the curve has no stationary points.

It looks like this:

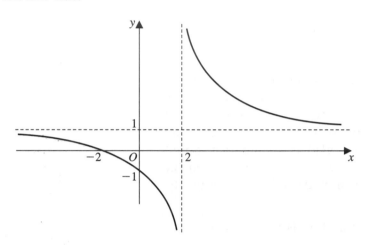

Example 7

Sketch the curve with equation $y = \dfrac{x}{4-x^2}$.

(1) When $x = 0$, $y = 0$

So the curve passes through the origin.

(2) When $x = \pm 2$, $4 - x^2 = 0$ and so y is infinite. Thus $x = \pm 2$ are asymptotes.

(3) As $x \to +\infty$, $y \to 0$

$\qquad$ As $x \to -\infty$, $y \to 0$

(4) The function is odd and so the graph has rotational symmetry about the origin.

(5) The graph exists for all x except $x = \pm 2$.

(6) $\dfrac{\mathrm{d}y}{\mathrm{d}x} = \dfrac{(4-x^2)\cdot 1 - x(-2x)}{(4-x^2)^2}$

$\qquad = \dfrac{x^2 + 4}{(4-x^2)^2}$

$\dfrac{\mathrm{d}y}{\mathrm{d}x} = 0 \Rightarrow x^2 + 4 = 0 \Rightarrow x^2 = -4$

No real value of x exists such that $x^2 = -4$ so the curve has no stationary point. It looks like this:

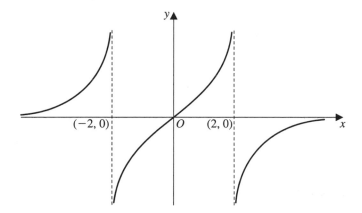

Example 8

Sketch the graph of $y = \dfrac{x^2 + 9}{x^2 - 1}$

(1) At $x = 0$, $y = \dfrac{9}{-1} = -9$

The curve cuts the y-axis at $y = -9$.

At $y = 0$, $x^2 + 9 = 0 \Rightarrow x^2 = -9$

No real values of x exist such that $x^2 = -9$. So the graph does not cut the x-axis.

(2) At $x = 1$ or $x = -1$, $x^2 - 1 = 0$

Thus y is infinite for these values of x.

Therefore the lines $x = -1$ and $x = 1$ are asymptotes.

(3) $\dfrac{x^2 + 9}{x^2 - 1} = \dfrac{x^2 - 1 + 10}{x^2 - 1} = 1 + \dfrac{10}{x^2 - 1}$

So as $x \to \pm\infty$, $\dfrac{10}{x^2 - 1} \to 0$ and $y \to 1$

So $y = 1$ is an asymptote.

(4) $\dfrac{x^2 + 9}{x^2 - 1}$ is an even function, so the graph is symmetrical about the y-axis.

(5) $y = \dfrac{x^2 + 9}{x^2 - 1}$ is defined for all x except $x = \pm 1$.

(6) $\dfrac{dy}{dx} = \dfrac{(x^2 - 1)2x - (x^2 + 9)2x}{(x^2 - 1)^2}$

$\dfrac{dy}{dx} = 0 \Rightarrow -2x - 18x = 0$

$\Rightarrow x = 0$

Since $\dfrac{dy}{dx} = \dfrac{-20x}{(x^2 - 1)^2}$,

when x is just negative, $\dfrac{dy}{dx} > 0$

When x is just positive, $\dfrac{dy}{dx} < 0$

So $(0, -9)$ is a maximum.

The curve with equation $y = \dfrac{x^2 + 9}{x^2 - 1}$ looks like this:

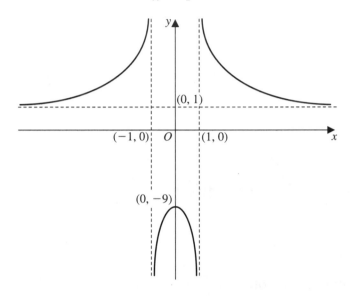

Sketching the graph of $y = \dfrac{1}{f(x)}$

If you are given, or are able to draw for yourself, a sketch of the graph of $y = f(x)$, then it is relatively simple to sketch the graph of $y = \dfrac{1}{f(x)}$ if you remember the following:

(1) If at $x = a$, the value of $f(x)$ is b, $b \neq 0$, then at $x = a$, the value of $\dfrac{1}{f(x)} = \dfrac{1}{b}$.

(2) If, at $x = a$, the value of $f(x)$ is 0, then as $x \to a$, $\dfrac{1}{f(x)} \to \infty$; that is, for the graph of $y = \dfrac{1}{f(x)}$, $x = a$ is an asymptote.

(3) If $x = a$ is an asymptote to the graph of $y = f(x)$, then, at $x = a$, $\dfrac{1}{f(x)} = 0$.

(4) If at $x = a$, $f(x) > 0$, then $\dfrac{1}{f(x)} > 0$ and if $f(x) < 0$ then $\dfrac{1}{f(x)} < 0$.

(5) If at $x = a$, there is a local maximum on the graph of $y = f(x)$, then at $x = a$ there is a local minimum on the graph of $y = \dfrac{1}{f(x)}$, and vice versa.

Example 9

On the same axes sketch the graphs of $y = (x + 1)(x - 2)$ and $y = \dfrac{1}{(x + 1)(x - 2)}$.

The graph of $y = (x + 1)(x - 2)$ is a parabola of the 'valley' variety which cuts the *x*-axis at $x = -1$ and $x = 2$ and cuts the *y*-axis at $y = -2$. So its graph looks like this:

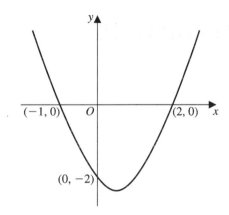

The value of y is 0 at $x = -1$ and at $x = 2$; so the graph of $y = \dfrac{1}{(x + 1)(x - 2)}$ has asymptotes at $x = -1$ and at $x = 2$.

The graph of $y = (x + 1)(x - 2)$ cuts the *y*-axis at $y = -2$; so the graph of $y = \dfrac{1}{(x + 1)(x - 2)}$ cuts the *y*-axis at $y = -\frac{1}{2}$.

The graph of $y = (x + 1)(x - 2)$ has a minimum value of $-\frac{9}{4}$ at $x = \frac{1}{2}$; so the graph of $y = \dfrac{1}{(x + 1)(x - 2)}$ has a maximum value of $-\frac{4}{9}$ at $x = \frac{1}{2}$.

On the graph of $y = (x + 1)(x - 2)$ as $x \to \pm\infty$, $y \to +\infty$; so on the graph of $y = \dfrac{1}{(x + 1)(x - 2)}$, as $x \to \pm\infty$, $y \to 0$.

The graph of $y = \dfrac{1}{(x+1)(x-2)}$ looks like this:

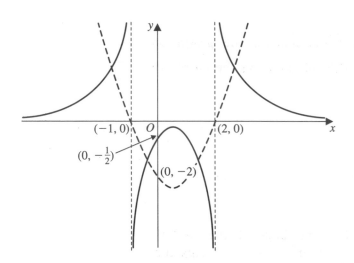

Exercise 3B

Sketch the curves with the following equations:

1 $y = x^3 - 3x^2$

2 $y = x^4 - 9x^2$

3 $y = \dfrac{1}{x+3}$

4 $y = \dfrac{x}{x-2}$

5 $y = \dfrac{x+3}{x+5}$

6 $y = \dfrac{3x-1}{x+1}$

7 $y = \dfrac{1}{x^2-16}$

8 $y = \dfrac{x^2}{x^2-16}$

9 $y = x^2 - x - 2$

10 $y = \dfrac{1}{x^2-x-2}$

11 $y = \dfrac{x}{(x+2)(x-3)}$

12 $y = \dfrac{(x+2)(x-3)}{x}$

13 $y = (2x-3)(x+2)$

14 $y = \dfrac{1}{(2x-3)(x+2)}$

15 $y = \dfrac{1}{(x+2)(3-x)}$

16 $y = (x+4)^3$

17 $y = x(x-5)^2$

18 $y = (4-x^2)(3+x)$

19 $y = |x|(x-5)$

20 $y = |(4-x^2)(3+x)|$

3.3 Sketching curves given by parametric equations

The coordinates of a point on a curve can often be expressed in terms of a third variable, called a parameter. For example, the curve with equation $y = 2x^2$ has parametric equations $x = t$, $y = 2t^2$ and any point P on the curve can be represented by $P(t, 2t^2)$.

The parametric equations of a curve are generally used when they are of a simpler form than the cartesian equation of the curve. Under these circumstances the mathematics involved in working with the parametric equations becomes much easier to handle.

Parametric equations that easily transform to a cartesian equation

It is often possible to transform the parametric equations of a curve into the corresponding cartesian equation. This is done by eliminating the parameter, t, say, between the parametric equations $x = f(t)$ and $y = g(t)$.

For example, if $x = 2t^2$ and $y = 4t$ then $t^2 = \dfrac{x}{2}$ and $t = \dfrac{y}{4}$.

So:
$$\frac{x}{2} = \left(\frac{y}{4}\right)^2$$

That is:
$$\frac{x}{2} = \frac{y^2}{16}$$

or
$$y^2 = 8x$$

The technique is to take all the t's (or whatever the parameter is) to the left-hand side of each of the parametric equations and everything else to the right-hand side. The two equations can then be linked as in the example above. When $t^2 = \dfrac{x}{2}$ and $t = \dfrac{y}{4}$, then also $t^2 = \left(\dfrac{y}{4}\right)^2$. So if $t^2 = \dfrac{x}{2}$ and $t^2 = \left(\dfrac{y}{4}\right)^2$ then

$$\frac{x}{2} = \left(\frac{y}{4}\right)^2 = \frac{y^2}{16} \Rightarrow y^2 = 8x$$

It is of the utmost importance when you do this that you make sure that the rearranged equation in x has no parameter on the right-hand side and, similarly, that the rearranged equation in y has no parameter on the right-hand side. Unless you do this you will find that your 'cartesian' equation will still involve a parameter. So you will have an equation with *three* variables – x, y and a parameter!

If you can transform the parametric equations into the cartesian equation you will often find that from the cartesian equation you can sketch the graph using the techniques you have already been shown.

Example 10

Sketch the curve given by the parametric equations

$$x = 3t, \ y = 1 + t^2$$

From the first equation:

$$t = \frac{x}{3}$$

From the second equation:

$$t^2 = y - 1$$

So:

$$y - 1 = \left(\frac{x}{3}\right)^2$$

i.e.

$$y = \frac{x^2}{9} + 1$$

This represents a parabola with a 'valley'. If the equation of the curve were $y = \frac{x^2}{9}$, the curve would have its minimum at the origin. So the graph of $y = \frac{x^2}{9} + 1$ has its minimum at $(0, +1)$, using the result found in section 2.7 of Book P2.

The graph looks like this:

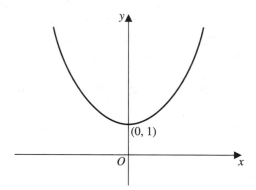

Example 11

Sketch the curve with parametric equations

$$x = t + 2, \ y = \frac{1}{t}$$

From the first equation $t = x - 2$. From the second equation $t = \frac{1}{y}$.

So:

$$x - 2 = \frac{1}{y}$$

or

$$y = \frac{1}{x - 2}$$

Consider the equation $y = \frac{1}{x}$.

As $x \to 0$, $y \to \infty$

As $x \to +\infty$, $y \to 0$

As $x \to -\infty$, $y \to 0$

$$\frac{\mathrm{d}y}{\mathrm{d}x} = -\frac{1}{x^2}$$

This can never be zero so the function has no stationary points. The graph looks like this:

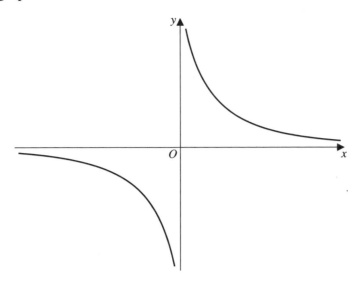

Now if $\mathrm{f}(x) \equiv \dfrac{1}{x}$

then:
$$\mathrm{f}(x - 2) \equiv \dfrac{1}{x - 2}$$

So, using the results from section 2.7 of Book P2, you can see that the graph of $y = \dfrac{1}{x - 2}$ looks like this:

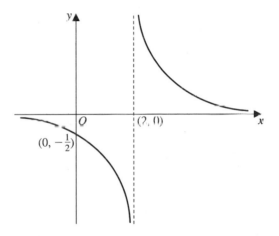

Curves that cannot easily be sketched from their cartesian equation or whose cartesian equation is difficult to obtain

It is always worth trying to rewrite the parametric equations of a curve to obtain the cartesian equation. However, frequently the cartesian equation turns out to be as complicated, if not more complicated, than the parametric equations of the curve. So, by going down this route, you often end up with a problem that is as complicated as, or more complicated than the original question. In these circumstances you need an alternative strategy. This is to try to sketch the curve directly from the parametric equations. Before you try to do this, you need to be sure that you can plot (rather than sketch) a curve given by parametric equations.

Example 12

Draw the curve given by the parametric equations

$$x = t^2, \ y = t^3$$

t	-4	-3	-2	-1	0	1	2	3	4
$x = t^2$	16	9	4	1	0	1	4	9	16
$y = t^3$	-64	-27	-8	-1	0	1	8	27	64

The curve is called a semi-cubical parabola and looks like this:

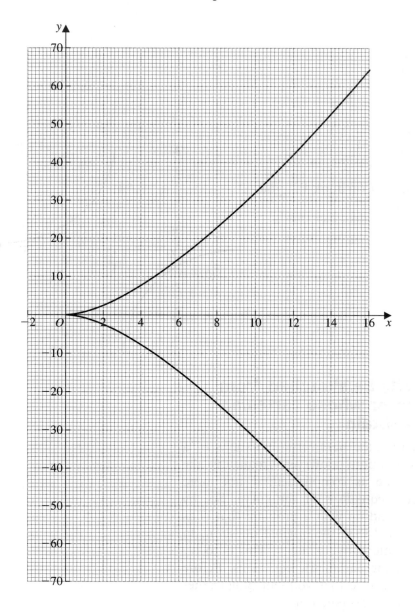

Example 13

Sketch the curve given by the parametric equations

$$x = t^2, \quad y = t^3 - t$$

(1) The curve cuts the x-axis where $y = 0$

i.e. where: $\qquad\qquad\qquad t^3 - t = 0$

$\Rightarrow \qquad\qquad\qquad t(t^2 - 1) = 0$

$\Rightarrow \qquad\qquad\qquad t(t - 1)(t + 1) = 0$

So: $\qquad\qquad\qquad t = 0, 1, -1$

For $t = 0$, $x = 0$. That is, the curve passes through $(0, 0)$.

For $t = \pm 1$, $x = 1$. That is, the curve passes through $(1, 0)$.

The curve cuts the y-axis where $x = 0$; that is where

$$t^2 = 0 \Rightarrow y = 0$$

So it cuts the y-axis only at $(0, 0)$.

(2) If $t = a$ (where a is positive), then:

$$x = a^2 \text{ and } y = a^3 - a$$

If $t = -a$, then:

$$x = a^2 \text{ (again) but } y = -a^3 + a = -(a^3 - a)$$

So the curve passes through $(a^2, a^3 - a)$ and through $(a^2, -a^3 + a)$. That is, the curve is symmetrical about the x-axis.

(3) Since for any real value of t, t^2 is positive, and since $x = t^2$, x can only be positive or zero. So there is no curve to the left of the y-axis.

Also, when t is very large $|t^3|$ is much larger than $|t|$. So:

as $t \to +\infty$, $x \to +\infty$ and $y \to +\infty$

and as $t \to -\infty$, $x \to +\infty$ and $y \to -\infty$

The graph therefore looks like this:

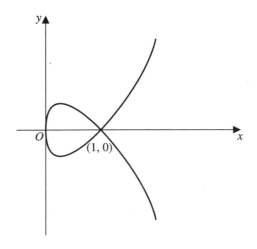

Curves with trigonometric parametric equations

Using parametric coordinates and parametric equations can often considerably simplify the algebra needed to derive the properties of plane curves.

■ **For the circle with centre O and radius a, the parametric equations $x = a\cos t$, $y = a\sin t$ are appropriate.**

The parameter t is restricted to the interval $0 \leqslant t < 2\pi$ so that there is one value of t corresponding to each point (x, y) on the curve, and for each (x, y) on the curve there is only one value of t.

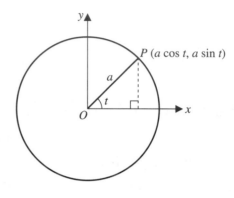

The parametric equations $x = a\cos t$, $y = a\sin t$ give you

$$\cos t = \frac{x}{a} \quad \text{and} \quad \sin t = \frac{y}{a}$$

Since $\cos^2 t + \sin^2 t \equiv 1$, then:

$$\left(\frac{x}{a}\right)^2 + \left(\frac{y}{a}\right)^2 = 1$$

That is, $x^2 + y^2 = a^2$ and the point P $(a\cos t, a\sin t)$ always lies on the circle with centre O and radius a.

■ **For the circle with equation $(x - \alpha)^2 + (y - \beta)^2 = r^2$ whose centre is at (α, β) and whose radius is r, suitable parametric equations are**
$$x = \alpha + r\cos\theta, \qquad y = \beta + r\sin\theta$$
where the parameter θ lies in the interval $0 \leqslant \theta < 2\pi$.

Note that t is often used as the parameter instead of θ.

Example 14

A circle is given by the equations
$$x = a + 2a\cos\theta, \quad y = 2a\sin\theta, \quad 0 \leqslant \theta < 2\pi$$

where a is a positive constant.

Sketch the circle, giving its cartesian equation.

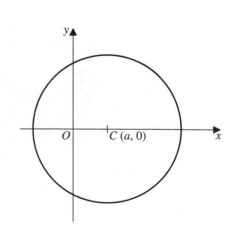

(a) Since $\cos\theta = \dfrac{x-a}{2a}$ and $\sin\theta = \dfrac{y}{2a}$, using the identity $\cos^2\theta + \sin^2\theta \equiv 1$ gives you

$$\frac{(x-a)^2}{4a^2} + \frac{y^2}{4a^2} = 1$$

That is, $(x - a)^2 + y^2 = 4a^2$ is the cartesian equation of the circle. So the centre of the circle, C, is at $(a, 0)$ and the radius of the circle is $2a$.

A sketch of the circle is shown opposite.

Example 15

Sketch the curve given by the equations

$$x = 3 \sin \theta, \ y = \cos \theta, \ 0 \leqslant \theta < 2\pi$$

Notice that $|x| \leqslant 3$, $|y| \leqslant 1$ because $\sin \theta$, $\cos \theta$ lie in $[-1, 1]$ and since $\sin^2 \theta + \cos^2 \theta \equiv 1$,

$$\left(\frac{x}{3}\right)^2 + y^2 = 1 \text{ is the cartesian equation}$$

That is:
$$\frac{x^2}{9} + y^2 = 1$$

This curve meets the x-axis at $(\pm 3, 0)$ and the y-axis at $(0, \pm 1)$. The curve is called an **ellipse** and a sketch of it is shown below.

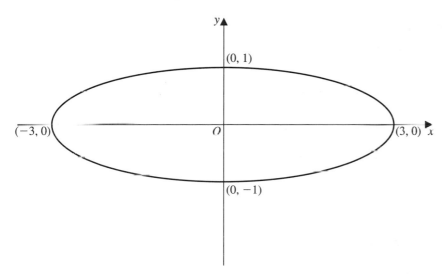

Exercise 3C

Sketch the curves given by the parametric equations:

1 $x = t^2 + 1, \ y = t$

2 $x = 1 + t, \ y = 4 - t^2$

3 $x = 4t, \ y = \dfrac{4}{t}$

4 $x = t - 1, \ y = t^2$

5 $x = 3 \cos \theta, \ y = 2 \sin \theta, \ 0 \leqslant \theta < 2\pi$

6 $x = 2 \sec t, \ y = 3 \tan t, \ -\dfrac{\pi}{2} < t < \dfrac{\pi}{2}$

7 $x = 2t + 3, \ y = t^2 - 1$

8 $x = 2t + 3, \ y = (t + 2)(t - 3)$

9 $x = \theta - \sin \theta, \ y = 1 - \cos \theta, \ 0 \leqslant \theta < 2\pi$

10 $x = \sin\theta, y = \cos 2\theta, 0 < \theta < \dfrac{\pi}{2}$

11 $x = 2\cos\theta, y = 2\sin\theta, 0 \leqslant \theta < 2\pi$

12 $x = 4\cos\theta, y = 3\sin\theta, 0 \leqslant \theta < 2\pi$

13 $x = \sin\theta, y = \sin 2\theta, 0 \leqslant \theta < 2\pi$

14 $x = t^2 - t, y = 2t$

15 $x = 2t^3, y = 3t^2$

16 $x = t^2 - 2, y = t^3 - 6t$

17 $x = t^2, y = t^3$

18 $x = t + 1, y = t^2 - 1$

19 $x = \dfrac{1}{t}, y = t + 3$

20 $x = t^2 - 1, y = t^4 + 1$

21 A circle is given parametrically by the equations
$$x = a(1 + \cos t), \quad y = a(3 + \sin t), \quad 0 \leqslant t < 2\pi$$
where a is a positive constant.

(a) Find the radius of the circle and the coordinates of its centre.

(b) Sketch the circle.

22 Find the radius and the coordinates of the centre of the circle given parametrically by the equations
$$x = 3\cos t + 2, \quad y = 3\sin t - 4, \quad 0 \leqslant t < 2\pi$$

23 A circle passes through the points (2, 5), (0, 1) and (0, 4). Prove that this circle touches the *x*-axis.

24 A point P on the circle with equation $x^2 + y^2 = a^2$ has coordinates $(a\cos\theta, a\sin\theta)$ and A is the point with coordinates $(2a, 4a)$.

(a) Write down the coordinates of M, the mid-point of AP.

(b) Given that θ may vary, find the cartesian equation of the locus of M.

SUMMARY OF KEY POINTS

1 The circle, centre the origin, radius a, has equation
$$x^2 + y^2 = a^2$$

2 The circle, centre (α, β), radius r has equation
$$(x - \alpha)^2 + (y - \beta)^2 = r^2$$

3 $x^2 + y^2 + 2gx + 2fy + c = 0$ is the equation of a circle, centre $(-g, -f)$ with radius $\sqrt{(g^2 + f^2 - c)}$.

4 The tangent at (h, k) to the circle with equation $x^2 + y^2 = a^2$ has equation

$$hx + ky = a^2$$

5 The tangent at (h, k) to the circle with equation $x^2 + y^2 + 2gx + 2fy + c = 0$ has equation

$$hx + ky + g(h + x) + f(k + y) + c = 0$$

6 Before attempting to sketch a curve given by a cartesian equation you should consider the following:
(a) Where does the curve cut the axes?
(b) Where are the asymptotes, if any?
(c) What happens as $x \to \pm \infty$?
(d) Has the curve any symmetry?
(e) Are there any points at which the curve is undefined?
(f) Where do the stationary points occur?

7 If you are asked to sketch a curve given by parametric equations, try to eliminate the parameter between $x = f(t)$ and $y = g(t)$ to obtain the cartesian equation of the curve and then proceed as in 6.

8 If you are asked to sketch a curve given by parametric equations and you either cannot obtain the cartesian equation or the cartesian equation is more complicated than the parametric equations, then sketch the curve from the parametric equations, asking the following questions:
(a) Where does the curve cut the axes?
(b) Does the curve have any symmetry?
(c) Are there any points at which the curve is undefined?
If all else fails, plot a few points.

9 For the circle with centre the origin and radius a, suitable parametric equations are

$$x = a \cos t, \qquad y = a \sin t, \qquad 0 \leqslant t < 2\pi$$

10 For the circle with centre (α, β) and radius r, suitable parametric equations are

$$x = \alpha + r \cos \theta, \qquad y = \beta + r \sin \theta, \qquad 0 \leqslant \theta < 2\pi.$$

Review exercise 1

1 Given that $e^{2x} + e^{2y} = xy$, find $\dfrac{dy}{dx}$ in terms of x and y. [E]

2 $$f(x) \equiv x^3 + ax^2 + bx + 6$$
Find, in terms of a and b, the remainder when $f(x)$ is divided by (a) $x - 2$ (b) $x + 3$.
Given that these remainders are equal, express a in terms of b.

3 Expand $(1 - 4x)^{\frac{1}{4}}$ in ascending powers of x up to and including the term in x^3, simplifying each coefficient.
State the set of values of x for which your series is valid.

4 Find the coordinates of the centre and the radius of the circle whose equation is

$$x^2 + y^2 - 16x - 12y + 96 = 0$$

Find also the least and the greatest distances of the origin O from the circumference of the circle.

5 (a) Given that $(x + 1)$ is a factor of the expression $(2x^3 + ax^2 - 5x - 2)$, find the value of the constant a.
Show that, with this value of a, $(x - 2)$ is another factor of this expression and hence, or otherwise, factorise the expression completely.
(b) When divided by $(x - 2)$ the expression $(x^3 + x^2 + 2x + 2)$ leaves a remainder R. Find the value of R.
[E]

6 The population, p, of insects on an island, t hours after midday, is given by

$$p = 1000e^{kt}$$

where k is a constant.

Given that when $t = 0$, the rate of change of the population with respect to time is 100 per hour,

(a) find k.

(b) Find also the population of insects on the island when $t = 6$. Give your answer to 3 significant figures. [E]

7 Differentiate with respect to x

(a) $\dfrac{\sin x}{e^x}$ (b) $\ln(1 + \tan^2 x)$ [E]

8 When a metal cube is heated, the length of each edge increases at the rate of $0.03\,\mathrm{cm\,s^{-1}}$. Find the rate of increase, in $\mathrm{cm^2\,s^{-1}}$, of the total surface area of the cube, when the length of each edge is $8\,\mathrm{cm}$. [E]

9 Given that $(x - 2)$ is a factor of $f(x)$, where

$$f(x) \equiv x^3 - x^2 + Ax + B$$

find an equation satisfied by the constants A and B.
Given, further, that when $f(x)$ is divided by $(x - 3)$ the remainder is 10, find a second equation satisfied by A and B. Solve your equations to find A and B.
Using your values of A and B, find 3 values of x for which $f(x) = 0$. [E]

10 Find the coordinates of the turning points on the curve with equation

$$y^3 + 3xy^2 - x^3 = 3$$ [E]

11 Find, in cartesian form, an equation of the circle which passes through $(3, 5)$, $(0, 4)$ and $(0, -4)$.

12 The radius of a circular ink blot is increasing at the rate of $0.3\,\mathrm{cm\,s^{-1}}$. Find, in $\mathrm{cm^2\,s^{-1}}$ to 2 significant figures, the rate at which the area of the blot is increasing at the instant when the radius of the blot is $0.8\,\mathrm{cm}$. [E]

13 $$f(x) \equiv \frac{2}{(2 - x)(1 + x)^2}$$

Express $f(x)$ in the form

$$\frac{A}{(2 - x)} + \frac{B}{(1 + x)} + \frac{C}{(1 + x)^2}$$

where A, B and C are numbers to be found. [E]

14 Express $f(x) \equiv \dfrac{5}{x^2 + x - 6} - \dfrac{1}{x^2 + 5x + 6}$ in partial fractions in its simplest form.

Given that $|x| < 2$, expand $f(x)$ in ascending powers of x as far as the term in x^2.

15 (a) Differentiate with respect to x
(i) $\ln(x^2)$
(ii) $x^2 \sin 3x$
(b) Find the gradient of the curve with equation

$$5x^2 + 5y^2 - 6xy = 13$$

at the point $(1, 2)$. [E]

16 Differentiate $e^{2x} \cos x$ with respect to x.
The curve C has equation $y = e^{2x} \cos x$.
(a) Show that the turning points on C occur where $\tan x = 2$.
(b) Find an equation of the tangent to C at the point where $x = 0$. [E]

17 Find the value of each of the constants A, B and C for which

$$\frac{1}{1 + x^3} \equiv \frac{A}{1 + x} + \frac{Bx + C}{1 - x + x^2}$$ [E]

18 Find the centre and the radius of each of the circles with equations

$$x^2 + y^2 - 8x - 6y = 0$$
$$x^2 + y^2 - 24x - 18y + 200 = 0$$

Deduce that the circles touch each other externally. [E]

19 Evaluate $\dfrac{dy}{dx}$

(a) at $\left(\frac{3}{4}\pi, 0\right)$ when $y = \ln(1 + \cos 2x)$
(b) at $(1, 2)$ when $x^2 y + y^2 = 6$ [E]

20 Show that the volume V of a right circular cone of slant height l and semi-vertical angle θ is given by

$$V = \tfrac{1}{3}\pi l^3 \sin^2 \theta \cos \theta$$

Find the maximum value of V for fixed l as θ varies between 0 and $\frac{1}{2}\pi$. [E]

21 Express $\dfrac{1+2x}{(1-3x)(1+6x^2)}$ in partial fractions.

Hence expand $\dfrac{1+2x}{(1-3x)(1+6x^2)}$ in ascending powers of x as far as the term in x^3. State the set of values of x for which the expansion is valid.

22 Find $\dfrac{dy}{dx}$ if

(a) $y = \dfrac{\sin 2x}{\cos 3x}$ (b) $y = \tan^5 x$ (c) $x^2 + y^3 = 12$

Find an equation of the tangent to the curve in (c) at the point $(2, 2)$.

23 (a) Differentiate the following functions with respect to x, simplifying your answers where possible:

(i) $\dfrac{\sqrt{(1+x^3)}}{x^2}$ (ii) $\ln\left(\dfrac{2+\cos x}{3-\sin x}\right)$

(b) If $y = e^{3x}\sin 4x$ show that

$$\frac{d^2y}{dx^2} - 6\frac{dy}{dx} + 25y = 0 \qquad\qquad \text{[E]}$$

24 The edges of a cube are of length x cm. Given that the volume of the cube is being increased at a rate of $p\,\text{cm}^3\,\text{s}^{-1}$, where p is a constant, calculate, in terms of p, in $\text{cm}^2\,\text{s}^{-1}$, the rate at which the surface area of the cube is increasing when $x = 5$.   [E]

25 Sketch the curves given parametrically by these equations, using a separate diagram for each.
(a) $x = 3\cos t,\ y = 3\sin t,\ 0 \leqslant t < 2\pi$
(b) $x = 4\cos t,\ y = 2\sin t,\ 0 \leqslant t < 2\pi$

(c) $x = 3\sec t,\ y = 3\tan t,\ -\dfrac{\pi}{3} \leqslant t \leqslant \dfrac{\pi}{3}$

Find a cartesian equation for each of these curves.

26 Show that $(1+x)(1+x^2) \equiv 1 + x + x^2 + x^3$.

Express $\dfrac{7+x}{1+x+x^2+x^3}$ in partial fractions.

Assuming that $|x| < 1$, expand $\dfrac{7+x}{1+x+x^2+x^3}$ in ascending powers of x as far as the term in x^4.

27 Given that $x^3 + y^3 = 6xy$, prove that

$$\frac{dy}{dx} = \frac{x^2 - 2y}{2x - y^2}$$

Hence find an equation of the normal to the curve with equation $x^3 + y^3 = 6xy$ at $(3, 3)$.

28 Given that $y = \sin^3 x \cos 3x$, prove that

$$\frac{dy}{dx} = 3 \sin^2 x \cos 4x$$ [E]

29 Sketch, on separate axes, the curves with equations

(a) $y = x^2(x - 2)$

(b) $y = \dfrac{1}{x^2(x - 2)}$

(c) $y^2 = x^2(x - 2)$

30 Express $\dfrac{3x - 5}{(x + 3)(2x - 1)}$ in partial fractions.

Find the first three terms in the series expansion of

$\dfrac{3x - 5}{(x + 3)(2x - 1)}$ in ascending powers of x and state the set of

values of x for which the expression is valid.

31 Find the coordinates of the centre and the radius of the circle with equation

$$x^2 + y^2 - 10y = 0$$

Prove that the line with equation $4x - 3y + 40 = 0$ is a tangent to the circle.

32 The curve shown has parametric equations

$$x = a \cos^3 t, \; y = a \sin^3 t$$

where $0 \leqslant t \leqslant \dfrac{\pi}{2}$ and a is a positive constant.

Prove that the tangent to the curve at the point P

where $t = \dfrac{\pi}{3}$ has equation $2x\sqrt{3} + 2y = a\sqrt{3}$.

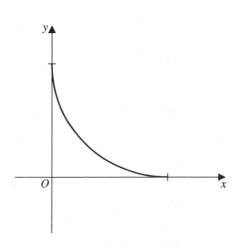

33 Find an equation of the circle whose centre is at $(-2, -5)$ and which passes through the point $(6, 1)$. Show that the tangent at $(6, 1)$ to the circle has equation $4x + 3y - 27 = 0$.

34 The number N of bacteria in a certain culture at time t hours
is given by $N = 600e^{ct}$, where c is a constant. Show that at
any instant the number of bacteria is increasing at a rate
proportional to the number of bacteria present at that instant.
The number of bacteria increases from 600 when $t = 0$ to
1800 when $t = 2$.
(i) Show that $c = \frac{1}{2}\ln 3$.
(ii) Show that the number of bacteria present at t hours is
$600B^{\frac{1}{2}t}$, where B is a constant and state the value of B.   [E]

35 (a) Given that $(x + 3)$ is a factor of the expression
$$5x^3 - px^2 - 53x + 84$$
(i) calculate the value of the constant p.
Hence, for this value of p,
(ii) factorise the expression completely.
(b) When the expression $4x^3 - 5x^2 + ax + 4$ is divided by
$(x - 2)$ there is a remainder of R. When the expression
$5x^3 + ax^2 + 6x + 4$ is divided by $(x - 2)$ there is a remainder
of $3R$. Find the value of the constant a.   [E]

36 Write y, where $y = \dfrac{x^2 + 8x + 9}{(x + 1)(x + 2)}$, as the sum of partial
fractions.
Find $\dfrac{dy}{dx}$ at $x = 1$.

37 A curve is given by the equations
$$x = t - 1, \ y = t^3$$
where t is a parameter.
(a) Sketch this curve.
(b) Find an equation of the normal to the curve at the point
where $t = -1$.

38 The rate of cooling of a metal ball placed in melting ice is
proportional to its own temperature $T\,°C$. Show that, at
time t,
$$T = Ae^{-kt}$$
where A and k are positive constants.
The temperature of the metal ball falls from $50\,°C$ to $40\,°C$ in
twenty seconds. Find its temperature at the end of the next
twenty seconds.   [E]

39
$$f(x) \equiv px^3 + 11x^2 + 2px - 5$$
The remainder when f(x) is divided by $(x+2)$ is 15.
(a) Show that $p = 2$.
(b) Factorise f(x) completely.
(c) Deduce the solutions of the equation
$$f(x) = (x+1)(x+5)$$ [E]

40 (a)
$$f(x) \equiv px^3 + qx^2 + rx + s$$
The curve with equation $y = $ f(x) has gradient 4 at the point with coordinates $(0,-5)$.
(i) Find the values of r and s.
The remainder when f(x) is divided by $(x-1)$ is 12, and the remainder when f(x) is divided by $(x+2)$ is 15.
(ii) Calculate the values of p and q.

(b)
$$g(x) \equiv 2x^3 - x^2 - 23x - 20$$
(i) Show that $(x+1)$ is a factor of g(x).
(ii) Factorise g(x) completely.
(iii) Solve the equation g(x) = 0. [E]

41 A curve C_1 has equation $2y^2 = x$; another curve C_2 is given by the parametric equations $x = 4t$, $y = \dfrac{4}{t}$, $t \neq 0$. On the same diagram, sketch C_1 and C_2 and find the coordinates of their point of intersection. [E]

42 Given that the polynomial P(x) is divisible by $(x-a)^2$, show that P′(x) is divisible by $(x-a)$.
The polynomial $(x^4 + x^3 - 12x^2 + px + q)$ is divisible by $(x+2)^2$. Find the values of the constants p and q. [E]

43 Carbon 14 is radioactive, and the count rate N, detected from the carbon 14 content of a piece of wood that is t years old, is given by
$$N = N_0 e^{-kt}$$
where N_0 is the count rate of new wood, and k is a positive constant.
Given that the count rate N of a piece of wood is halved after a period of 5600 years, calculate the value of k to 3 significant figures. [E]

44 Find the coordinates of the point of intersection of the curve with equation $xy = x^3 + 1$ and the x-axis. Find also the gradients at the points where $x = 1$ and $x = -1$. Find the value of x at the stationary point and the nature of the stationary point.

Sketch, on the same diagram, the curves with equations

$$y = \frac{x^3 + 1}{x} \text{ and } y = \frac{x}{x^3 + 1}$$ [E]

45 Given that $(x + 3)$ and $(x - 1)$ are both factors of the expression $(x^3 + ax^2 - bx - a)$, calculate the values of the constants a and b.

With these values of a and b, find the value of the remainder when the expression is divided by $(x - 2)$. [E]

46 $$f(x) \equiv \frac{9 - 3x - 12x^2}{(1 - x)(1 + 2x)}$$

Given that $f(x) \equiv A + \dfrac{B}{1 - x} + \dfrac{C}{1 + 2x}$, find the values of the constants A, B and C. [E]

47 Given that $y = \sec x + \tan x$, prove that
$$\frac{dy}{dx} - y \tan x = 1$$

48 The function f is defined by

$$f : x \mapsto \frac{2x + 1}{x - 1}, \ x \in \mathbb{R}, \ x \neq 1$$

(a) Find $f^{-1}(x)$ in terms of x.

(b) On the same diagram, sketch the curves with equations $y = f(x)$ and $y = f^{-1}(x)$.

49 Given that $y = 2 - \sin x + \frac{1}{2} \cos 2x$, $0 \leqslant x \leqslant 2\pi$, find the maximum and the minimum values of y and distinguish between them. [E]

50 (a) Given that $(x + 4)$ is a factor of the expression

$$3x^3 + x^2 + px + 24$$

find the value of the constant p.

Hence, factorise the expression completely.

(b) When divided by $(x + 2)$ the expression $(5x^3 - 3x^2 + ax + 7)$ leaves a remainder of R. When the expression $(4x^3 + ax^2 + 7x - 4)$ is divided by $(x + 2)$ there is a remainder of $2R$. Find the value of the constant a. [E]

51 Sketch the curve with equation $3y^2 = x^2(2x+1)$ and prove that the two tangents to the curve at the origin intersect at an angle of $\dfrac{\pi}{3}$.   [E]

52 Sketch the curve with parametric equations
$$3x = t^3 - 3t, \ 3y = t^3$$
Find an equation of the tangent to the curve at the point where $t = 3$.

53 Given that $f(x) \equiv 2x^3 - 3x^2 - 11x + c$, and that the remainder when $f(x)$ is divided by $x - 2$ is -12, find the value of the constant c.

Show that, with this value of c, $x + 2$ is a factor of $f(x)$, and hence solve the equation $f(x) = 0$.   [E]

54 For the curve with equation $y = \dfrac{x^2 + 1}{x^2 - 4}$ find

(a) the coordinates of the turning point

(b) the equations of the asymptotes.

Sketch the curves with equations $y = \dfrac{x^2 + 1}{x^2 - 4}$ and $y = \dfrac{x^2 - 4}{x^2 + 1}$ on the same axes.

55 (a) Given that $y = 18x - 16\sin x + \sin 2x$, prove that $\dfrac{dy}{dx} > 0$ for all values of x.

(b) Given that $y = x\sin 3x$, prove that
$$\frac{d^2y}{dx^2} + 9y = 6\cos 3x$$ [E]

56 Write down and simplify the expansion of $(1 - 8x)^{\frac{1}{2}}$, in ascending powers of x, up to and including the term in x^3.
State the set of values of x for which the expansion is valid.   [E]

57 $f(n) \equiv n^2 + n + 1$, where n is a positive integer.

Classify the following statements about $f(n)$ as true or false. If a statement is true, prove it; if it is false, provide a counter-example.

(a) $f(n)$ is always a prime number

(b) $f(n)$ is always an odd number.   [E]

58
$$f(x) \equiv \frac{2 - x}{(1 + 2x)(1 + x^2)}$$

Given that $f(x) \equiv \dfrac{A}{1 + 2x} + \dfrac{Bx + C}{1 + x^2}$,

(a) find the values of the constants A, B and C

(b) find $f'(x)$. You need not simplify your answer.   [E]

59 Differentiate, with respect to x,

(a) $\dfrac{\sin x}{x}$, $x > 0$

(b) $\ln\left(\dfrac{1}{x^2 + 9}\right)$.

Given that $y = x^x$, $x > 0$, $y > 0$, by taking logarithms

(c) show that $\dfrac{dy}{dx} = x^x(1 + \ln x)$.   [E]

60 Expand $(1 + x - 3x^2)^4$ in ascending powers of x as far as the term in x^3, showing that the coefficients of x and x^2 are 4 and -6 respectively.

If $a > b$ and the first three terms in the expansion, in ascending powers of x, of
$$\frac{1 + ax}{\sqrt{(1 + bx)}}$$

are the same as those in the previous expansion, find a and b. State the set of values of x for which this second expansion is valid.   [E]

61 A curve, C, is given by
$$x = 2t + 3, \; y = t^3 - 4t$$

where t is a parameter. The point A has parameter $t = -1$ and the line l is the tangent to C at A. The line l also intersects the curve at B.

(a) Show that an equation for l is $2y + x = 7$.

(b) Find the value of t at B.   [E]

62
$$f(x) \equiv e^{2x} \sin 2x, \; 0 \leqslant x \leqslant \pi$$

(a) Find the values of x for which $f(x) = 0$, giving your answers in terms of π.

(b) Use calculus to find the coordinates of the turning points on the graph of $y = f(x)$.

(c) Show that $f''(x) = 8e^{2x} \cos 2x$.

(d) Hence, or otherwise, determine which turning point is a maximum and which is a minimum.   [E]

63 (a) Find in cartesian form an equation of the circle C with centre $(1, 4)$ and radius 3.

(b) Determine, by calculation, whether the point $(2.9, 1.7)$ lies inside or outside C. [E]

64 The binomial expansion of $(8 + x)^{\frac{1}{3}}$ in ascending powers of x, as far as the term in x^2, is

$$(8 + x)^{\frac{1}{3}} = 2 + px + qx^2 + \cdots, |x| < 8$$

(a) Determine the values of the constants p and q.

(b) Use the expression $2 + px + qx^2$, and your values of p and q, to obtain an estimate for $\sqrt[3]{15}$, giving your answer to 3 significant figures.

(c) Find the percentage error involved in using this estimate. [E]

65 The rate of decay of a radioactive substance S is proportional to the amount remaining so that

$$\frac{dx}{dt} = -kx$$

where x is the amount of S present at time t and k is a positive constant. Given that $x = a$ at time $t = 0$, show that, if the time taken for the amount of S to become $\frac{1}{2}a$ is T, then

$$x = a\left(\tfrac{1}{2}\right)^{\frac{t}{T}}$$ [E]

66 The curve C has parametric equations

$$x = t^3, y = t^2, t > 0$$

(a) Find an equation of the tangent to C at $A(1, 1)$.

Given that the line l with equation $3y - 2x + 4 = 0$ cuts the curve C at point B,

(b) find the coordinates of B

(c) prove that the line l only cuts C at the point B. [E]

67 $f(x) \equiv 6x^3 + Ax^2 + x - 2$, where A is a constant. Given that $(x + 2)$ is a factor of $f(x)$,

(a) find the value of A.

Using this value of A,

(b) express $\dfrac{1}{f(x)}$ in partial fractions

(c) find the value of $\dfrac{d^2}{dx^2}\left[\dfrac{1}{f(x)}\right]$ at $x = 0$.

68 Given that $f(x) \equiv \dfrac{11 - 5x^2}{(x+1)(2-x)}$, find constants A and B such that

$$f(x) \equiv 5 + \frac{A}{x+1} + \frac{B}{2-x}$$

Given that x is so small that x^3 and higher powers of x may be neglected, find the series expansion of $f(x)$ in ascending powers of x up to and including the term in x^2.   [E]

69 A curve C is given by the equations

$$x = 2\cos t + \sin 2t, \; y = \cos t - 2\sin 2t, \; 0 \leqslant t < \pi,$$

where t is a parameter.

(a) Find $\dfrac{\mathrm{d}x}{\mathrm{d}t}$ and $\dfrac{\mathrm{d}y}{\mathrm{d}t}$ in terms of t.

(b) Find the value of $\dfrac{\mathrm{d}y}{\mathrm{d}x}$ at the point P on C where $t = \dfrac{\pi}{4}$.

(c) Find an equation of the normal to the curve at P.   [E]

70 The circle with equation $x^2 + y^2 - 6x + 8y + 9 = 0$ has centre C_1 and radius r_1.

The circle with equation $x^2 + y^2 + 4x - 16y - 13 = 0$ has centre C_2 and radius r_2.

(a) Find the coordinates of the points C_1 and C_2.

(b) Evaluate r_1 and r_2.

(c) Prove that the circles touch externally.

(d) Find an equation of the circle on $C_1 C_2$ as diameter.

71 Find $\dfrac{\mathrm{d}y}{\mathrm{d}x}$ in terms of x when

(a) $y = x^2 e^{-3x}$

(b) $y = \ln\left(\dfrac{1 + x^2}{1 - x^2}\right)$.

72 Expand $\ln\dfrac{(1 + 2x)^2}{(1 - 3x)}$ as a series in ascending powers of x as far as the term in x^3.   [E]

73
$$f(x) \equiv \frac{x + 4}{(x + 1)^2 (x + 2)}$$

(a) Express $f(x)$ in partial fractions.

(b) Find $f'(1)$.

74 Given that $y^2 = \sec x + \tan x$, prove that

(a) $\dfrac{dy}{dx} = \tfrac{1}{2} y \sec x$

(b) $\dfrac{d^2y}{dx^2} = \tfrac{1}{2} \sec x \left[\dfrac{dy}{dx} + y \tan x \right]$.

75 Find an equation of the tangent at $P\left(t, \dfrac{1}{t^2} \right)$ to the curve with

equation $y = \dfrac{1}{x^2}$. The tangent cuts the x-axis at Q and the

y-axis at R. Prove that

(a) $RP = 2PQ$ (b) $SR = PQ$

where S is the point on the curve where the tangent at P

meets the curve again.   [E]

76 (a) Expand $(1 - 3x)^{\frac{1}{3}}$, $|x| < \tfrac{1}{3}$, in ascending powers of x up to
and including the term in x^3.

(b) By substituting $x = 10^{-3}$ in your expansion, find, to 9
significant figures, the cube root of 997.   [E]

77 Find an equation of the circle which has the points $A(3, 1)$
and $B(-2, 2)$ at the ends of a diameter.

Given that A and B are opposite vertices of a square, find the
coordinates of the other two vertices.

78 Expand $(1 + x)^{\frac{1}{5}}$ in ascending powers of x as far as the term
in x^3.

Use your series with an appropriate value of x to find $33^{\frac{1}{5}}$
correct to 5 decimal places.   [E]

79 A spherical balloon has radius r metres at time t seconds and

$\dfrac{dr}{dt}$ is $k\,\mathrm{m\,s^{-1}}$, where k is a constant. Find, in terms of k, when

the radius is 2.5 metres, the rate of change of

(a) the surface area

(b) the volume.

80 Given that a curve has equation

$$y^2 + 3xy + 4x^2 = 37$$

find the value of $\dfrac{dy}{dx}$ at the point $(4, -3)$.

Integration

4

4.1 Integrating standard functions

In Books P1 and P2 you learned these standard integrals:

$$\int x^n \mathrm{d}x = \frac{1}{n+1} x^{n+1} + C, n \neq -1$$

$$\int e^x \mathrm{d}x = e^x + C$$

$$\int \frac{1}{x} \mathrm{d}x = \ln|x| + C$$

As integration is the reverse process to differentiation, you can employ the results obtained in chapter 2 to produce some more integrals. Here a, b are constants and C is the arbitrary constant of integration.

- $$\int \sin x \, \mathrm{d}x = -\cos x + C$$

- $$\int \sin(ax + b) \, \mathrm{d}x = -\frac{1}{a} \cos(ax + b) + C$$

- $$\int \cos x \, \mathrm{d}x = \sin x + C$$

- $$\int \cos(ax + b) \, \mathrm{d}x = \frac{1}{a} \sin(ax + b) + C$$

- $$\int (ax + b)^n \, \mathrm{d}x = \frac{1}{a(n+1)} (ax + b)^{n+1} + C, n \neq -1$$

- $$\int \frac{1}{ax + b} \, \mathrm{d}x = \frac{1}{a} \ln|ax + b| + C$$

- $$\int e^{ax+b} \, \mathrm{d}x = \frac{1}{a} e^{ax+b} + C$$

Often, memory plays a vital role in exercises involving integration. This is why **you should memorise these integrals and the results of chapter 2.**

Example 1

Find (a) $\displaystyle\int \operatorname{cosec}^2 x \, \mathrm{d}x$ (b) $\displaystyle\int \sec 2x \tan 2x \, \mathrm{d}x$

(a) In chapter 2, this result was found:

$$\frac{\mathrm{d}}{\mathrm{d}x}(\cot x) = -\operatorname{cosec}^2 x$$

By reversing this result, you get:

$$\int \operatorname{cosec}^2 x \, \mathrm{d}x = -\cot x + C$$

(b) In chapter 2 also, the result

$$\frac{\mathrm{d}}{\mathrm{d}x}(\sec x) = \sec x \tan x$$

was established, using the chain rule.

So:
$$\frac{\mathrm{d}}{\mathrm{d}x}(\sec 2x) = 2 \sec 2x \tan 2x$$

By reversing this result you get

$$\int \sec 2x \tan 2x \, \mathrm{d}x = \tfrac{1}{2} \sec 2x + C$$

Example 2

Evaluate $\displaystyle\int_0^{\frac{\pi}{3}} (\cos 3x - 2 \sin x) \, \mathrm{d}x$

You know that

$$\int \cos 3x \, \mathrm{d}x = \tfrac{1}{3} \sin 3x \quad \text{and} \quad \int \sin x \, \mathrm{d}x = -\cos x$$

So:
$$\int (\cos 3x - 2 \sin x) \, \mathrm{d}x = \tfrac{1}{3} \sin 3x + 2 \cos x + C$$

$$\left[\tfrac{1}{3} \sin 3x + 2 \cos x \right]_0^{\frac{\pi}{3}} = \tfrac{1}{3} \sin \pi + 2 \cos \tfrac{\pi}{3} - \left(\tfrac{1}{3} \sin 0 + 2 \cos 0 \right)$$
$$= 0 + 2(\tfrac{1}{2}) - 0 - 2(1)$$
$$= -1$$

Example 3

Evaluate $\displaystyle\int_1^6 \frac{1}{3x + 2} \, \mathrm{d}x$

We have

$$\int \frac{1}{3x + 2} \, \mathrm{d}x = \tfrac{1}{3} \ln|3x + 2| + C$$

from the list of standard integrals.

$$\int_1^6 \frac{1}{3x+2}\, dx = \left[\tfrac{1}{3} \ln |3x+2| \right]_1^6$$
$$= \tfrac{1}{3}\ln 20 - \tfrac{1}{3}\ln 5$$
$$= \tfrac{1}{3}\ln \tfrac{20}{5} = \tfrac{1}{3}\ln 4$$

Example 4

Find $\int (e^x + 1)^2\, dx$

$$\int (e^x + 1)^2\, dx = \int (e^{2x} + 2e^x + 1)\, dx$$
$$= \int e^{2x}\, dx + 2 \int e^x\, dx + \int 1\, dx$$
$$= \tfrac{1}{2} e^{2x} + 2e^x + x + C$$

Exercise 4A

Integrate with respect to x:

1 $\cos 4x$

2 $\sin 3x$

3 $\sin \tfrac{1}{2}x$

4 $\cos \dfrac{3x}{2}$

5 $\sec^2 x$

6 $\operatorname{cosec}^2 3x$

7 e^{-2x}

8 e^{3x-2}

9 $\dfrac{1}{2x-5}$

10 $(4x-3)^3$

11 $\cos(5x+4)$

12 $\sin(3-4x)$

13 $(3-2x)^2$

14 $\dfrac{1}{(3-2x)^2}$

15 $(e^x - e^{-x})^2$

16 $\dfrac{2}{\cos^2 \frac{x}{2}}$

17 $\sec 3x \tan 3x$

18 $\operatorname{cosec} 2x \cot 2x$

19 $\sec^2 x - x^2$

20 $2x + \sin 2x$

Evaluate:

21 $\displaystyle\int_{-\frac{\pi}{3}}^{\frac{\pi}{2}} \sin x\, dx$

22 $\displaystyle\int_{-\frac{\pi}{6}}^{\frac{\pi}{2}} \cos x\, dx$

23 $\displaystyle\int_{\frac{\pi}{3}}^{\frac{2\pi}{3}} \sin 2x\, dx$

24 $\displaystyle\int_0^{\frac{\pi}{4}} \sec^2 x\, dx$

25 $\displaystyle\int_0^{\frac{\pi}{2}} (x + \sin x)\, dx$

26 $\displaystyle\int_{\frac{\pi}{2}}^{\frac{\pi}{3}} \operatorname{cosec}^2 \tfrac{1}{2}x\, dx$

27 $\displaystyle\int_0^2 \frac{1}{3x+4}\, dx$

28 $\displaystyle\int_1^2 (2x-1)^3\, dx$

29 $\displaystyle\int_{-1}^1 e^{2x-1}\, dx$

30 $\displaystyle\int_0^{\frac{2\pi}{3}} \sec \tfrac{1}{2}x \, \tan \tfrac{1}{2}x\, dx$

4.2 Integration using identities

Until now you have been finding integrals by just thinking of integration as the reverse of differentiation. But often this approach does not work immediately because the function to be integrated is not integrable as it stands. In this case it is often possible to replace the function by equivalent ones using, for example, trigonometric identities, which can then be immediately integrated by using the formula

$$\int (u \pm v)\, dx = \int u\, dx \pm \int v\, dx.$$

The following examples illustrate this.

Example 5

Find $\displaystyle\int \tan^2 x\, dx$.

Consider the identity

$$\sec^2 x \equiv \tan^2 x + 1$$

which can be rearranged as

$$\tan^2 x \equiv \sec^2 x - 1$$

That is,

$$\int \tan^2 x\, dx = \int (\sec^2 x - 1)\, dx$$

$$= \int \sec^2 x\, dx - \int 1\, dx$$

So: $$\int \tan^2 x\, dx = \tan x - x + C$$

Note: We know that $\displaystyle\int \sec^2 x\, dx = \tan x + C$ because

$$\frac{d}{dx}(\tan x) = \sec^2 x$$

Example 6

Use the identity $\cos 2x \equiv 2\cos^2 x - 1$ to find $\displaystyle\int \cos^2 x\, dx$.

If you rearrange the identity,

$$\cos^2 x \equiv \tfrac{1}{2} + \tfrac{1}{2}\cos 2x$$

So:
$$\int \cos^2 x \, dx = \int (\tfrac{1}{2} + \tfrac{1}{2}\cos 2x) \, dx$$

$$= \int \tfrac{1}{2} \, dx + \int \tfrac{1}{2}\cos 2x \, dx$$

$$= \tfrac{1}{2}x + \tfrac{1}{4}\sin 2x + C$$

Another way to deal with an expression that you cannot immediately integrate is to split it into partial fractions that *can* be integrated.

Example 7

Find $\displaystyle\int \frac{7 - 5x}{(2x - 1)(x + 1)} \, dx$.

First you split the expression $\dfrac{7 - 5x}{(2x - 1)(x + 1)}$ into partial fractions by writing:

$$\frac{7 - 5x}{(2x - 1)(x + 1)} \equiv \frac{A}{2x - 1} + \frac{B}{x + 1}$$

So:
$$7 - 5x \equiv A(x + 1) + B(2x - 1)$$

Let $x = -1$, then $B = -4$

Let $x = \tfrac{1}{2}$, then $A = 3$

Then
$$\frac{7 - 5x}{(2x - 1)(x + 1)} \equiv \frac{3}{2x - 1} - \frac{4}{x + 1}$$

To find the required integral, write:

$$\int \frac{7 - 5x}{(2x - 1)(x + 1)} \, dx = \int \frac{3}{2x - 1} \, dx - \int \frac{4}{x + 1} \, dx$$

$$= \tfrac{3}{2}\ln|2x - 1| - 4\ln|x + 1| + C$$

Exercise 4B

Integrate with respect to x:

1 $\sin^2 x$ 2 $\cot^2 x$ 3 $\tan^2 2x$

4 $(1 + \cos x)^2$ 5 $(1 - 2\sin x)^2$ 6 $(\cos x + \sec x)^2$

7 $\dfrac{1}{x^2 - 4}$ 8 $\dfrac{4 - x}{(x - 2)(x - 3)}$ 9 $\dfrac{5 - 2x}{(x - 1)(2x + 1)}$

10 $\dfrac{x}{(2x + 1)(3x + 1)}$ 11 $\dfrac{-1}{(2x + 1)(3x + 1)}$

12 $\dfrac{12}{(3-2x)(3+2x)}$

13 By finding A, B and C so that

$$\frac{x^2}{x^2-1} \equiv A + \frac{B}{x-1} + \frac{C}{x+1}$$

find $\displaystyle\int \frac{x^2}{x^2-1}\, dx$.

14 Show that $\sin^2 x + 3\cos^2 x \equiv 2 + \cos 2x$.

Hence evaluate $\displaystyle\int_{\frac{\pi}{12}}^{\frac{\pi}{4}} (\sin^2 x + 3\cos^2 x)\, dx$.

15 Show that $\dfrac{4\cos 2x}{\sin^2 2x} \equiv \operatorname{cosec}^2 x - \sec^2 x$.

Hence evaluate $\displaystyle\int_{\frac{\pi}{6}}^{\frac{\pi}{3}} \frac{4\cos 2x}{\sin^2 2x}\, dx$.

16 Evaluate $\displaystyle\int_{0}^{\frac{\pi}{6}} (\sin 3x + \cos 2x)\, dx$.

17 Evaluate (a) $\displaystyle\int_{0}^{\pi} \sin^2 \tfrac{1}{4} x\, dx$ (b) $\displaystyle\int_{0}^{\pi} \cos^2 \tfrac{1}{4} x\, dx$.

4.3 Integration using substitutions

The integral $\displaystyle\int f(x)f'(x)\, dx$ is often called **the integral of a function f and its derivative f'.** By a simple substitution of $f(x) = u$, say, you can transform the integral into an integral which is simpler, with the variable u replacing the variable x.

$$u = f(x) \Rightarrow \frac{du}{dx} = f'(x)$$

So:
$$\int f(x)f'(x)\, dx = \int u \frac{du}{dx}\, dx$$

$$= \int u\, du$$

Integrating gives

$$\int f(x)f'(x)\, dx = \tfrac{1}{2}u^2 + C$$

That is:

■ $\int f(x)f'(x)\,dx = \frac{1}{2}[f(x)]^2 + C$

You should also note the general results

■ $\int [f(x)]^n f'(x)\,dx = \dfrac{1}{n+1}\,[f(x)]^{n+1} + C,\, n \neq -1$

■ $\int \dfrac{f'(x)}{f(x)}\,dx = \ln|f(x)| + C$

Once you are thoroughly practised in the techniques of integration, you will find that, in some simple cases, you can recognise a function and its derivative just by looking at the expression to be integrated. If you can do this, you can move straight to the answer without employing a change of variable. You are advised, however, in the early stages of this work to use the substitution method.

Example 8

Use the substitution $u = \cos x$ to find $\int \tan x\,dx$.

First you need to write

$$\int \tan x\,dx = \int \frac{\sin x}{\cos x}\,dx$$

Let $u = \cos x$, then

$$\frac{du}{dx} = -\sin x$$

and

$$\sin x = -\frac{du}{dx}$$

Then: $\int \dfrac{\sin x}{\cos x}\,dx = \int \dfrac{1}{u}\cdot\left(-\dfrac{du}{dx}\right)dx = -\int \dfrac{1}{u}\,du$

Integrating gives:

$$\int \frac{1}{u}\,du = \ln|u| + C$$

But $\int \tan x\,dx = -\int \dfrac{1}{u}\,du$ and $u = \cos x$.

So: $\int \tan x\,dx = -\ln|\cos x| + C$

But $\sec x = \dfrac{1}{\cos x}$ and

$$\ln \sec x = \ln \frac{1}{\cos x} = \ln 1 - \ln \cos x$$
$$= -\ln \cos x$$

- **So:**
$$\int \tan x \, dx = \ln |\sec x| + C$$

You can use a similar method to show that

-
$$\int \cot x \, dx = \ln |\sin x| + C$$

Example 9

Use the substitution $u^2 = x + 1$ to find

$$\int \frac{x}{(x+1)^{\frac{1}{2}}} \, dx$$

Differentiating $x + 1 = u^2$ gives $\dfrac{dx}{du} = 2u$;

also: $(x+1)^{\frac{1}{2}} = u$ and $x = u^2 - 1$

So:
$$\int \frac{x}{(x+1)^{\frac{1}{2}}} \, dx = \int \frac{u^2 - 1}{u} \, dx$$

You cannot integrate a function of u with respect to x and so you must replace the 'dx' by writing

$$dx = \frac{dx}{du} \cdot du$$

This technique is analogous to the chain rule in differentiation. It changes the variable of integration from x to u.

That is
$$\int \frac{x}{(x+1)^{\frac{1}{2}}} \, dx = \int \left(\frac{u^2 - 1}{u} \right) \left(\frac{dx}{du} \right) du$$

$$= \int \frac{u^2 - 1}{u} \, (2u) \, du$$

$$= \int (2u^2 - 2) \, du$$

Integrating with respect to u gives

$$\int \frac{x}{(x+1)^{\frac{1}{2}}} \, dx = \tfrac{2}{3} u^3 - 2u + C$$

Reverting to the variable x gives

$$\int \frac{x}{(x+1)^{\frac{1}{2}}} \, dx = \tfrac{2}{3}(x+1)^{\frac{3}{2}} - 2(x+1)^{\frac{1}{2}} + C$$

It is important to appreciate that you do not have to revert to the original variable when evaluating a *definite* integral. Suppose that you were required to evaluate

$$\int_0^3 \frac{x}{(x+1)^{\frac{1}{2}}} \, dx$$

As $u^2 = x + 1$, you can see that at $x = 0$, $u = 1$ and at $x = 3$, $u = 2$.

So :
$$\int_0^3 \frac{x}{(x+1)^{\frac{1}{2}}} \, dx = \int_1^2 (2u^2 - 2) \, du = \left[\tfrac{2}{3}u^3 - 2u \right]_1^2$$

$$= \tfrac{16}{3} - 4 - (\tfrac{2}{3} - 2)$$

$$= \tfrac{8}{3} = 2\tfrac{2}{3}$$

Example 10

Use the substitution $x = \operatorname{cosec} t$ to evaluate

$$\int_{\frac{2}{\sqrt{3}}}^2 \frac{1}{x^2 \sqrt{(x^2 - 1)}} \, dx$$

If $x = \operatorname{cosec} t$, then $\dfrac{dx}{dt} = \operatorname{cosec} t \cot t$ (see chapter 2)

and $\sqrt{(x^2 - 1)} = \sqrt{(\operatorname{cosec}^2 t - 1)} = \sqrt{(\cot^2 t)} = \cot t$

(from using the identity $\operatorname{cosec}^2 t \equiv 1 + \cot^2 t$).

At $x = 2$: $\operatorname{cosec} t = 2 \Rightarrow t = \dfrac{\pi}{6}$

At $x = \dfrac{2}{\sqrt{3}}$: $\operatorname{cosec} t = \dfrac{2}{\sqrt{3}} \Rightarrow t = \dfrac{\pi}{3}$

So:
$$\int \frac{1}{x^2 \sqrt{(x^2 - 1)}} \, dx = \int \frac{1}{x^2 \sqrt{(x^2 - 1)}} \frac{dx}{dt} \, dt$$

and
$$\int_{\frac{2}{\sqrt{3}}}^2 \frac{1}{x^2 \sqrt{(x^2 - 1)}} \, dx = \int_{\frac{\pi}{3}}^{\frac{\pi}{6}} \frac{1}{\operatorname{cosec}^2 t \cot t} (-\operatorname{cosec} t \cot t) \, dt$$

$$= -\int_{\frac{\pi}{3}}^{\frac{\pi}{6}} \frac{1}{\operatorname{cosec} t} \, dt = -\int_{\frac{\pi}{3}}^{\frac{\pi}{6}} \sin t \, dt$$

Integrating gives

$$\int_{\frac{2}{\sqrt{3}}}^2 \frac{1}{x^2 \sqrt{(x^2 - 1)}} \, dx = \left[\cos t \right]_{\frac{\pi}{3}}^{\frac{\pi}{6}} = \cos\tfrac{\pi}{6} - \cos\tfrac{\pi}{3}$$

$$= \frac{\sqrt{3}}{2} - \frac{1}{2} = \frac{\sqrt{3} - 1}{2}$$

Exercise 4C

In each of the following find the integral by using the substitution given. Your final result should be given in terms of x.

1 $\displaystyle\int \sin^3 x \cos x \, dx; \quad u = \sin x$

2 $\displaystyle\int \tan^2 x \sec^2 x \, dx; \quad u = \tan x$

3 $\displaystyle\int x(x^2 + 1)^3 \, dx; \quad u = x^2 + 1$

4 $\displaystyle\int x^3 \sqrt{(x^4 - 1)} \, dx; \quad u = x^4$

5 $\displaystyle\int \frac{e^x}{\sqrt{(e^x - 1)}} \, dx; \quad u = e^x$

6 $\displaystyle\int \sec^2 x \tan x \, dx; \quad u = \sec x$

7 $\displaystyle\int x e^{x^2} \, dx; \quad u = x^2$

8 $\displaystyle\int \frac{(\ln x)^2}{x} \, dx; \quad u = \ln x$

9 $\displaystyle\int \left(\frac{x}{x+1}\right)^2 dx; \quad u = x + 1$

10 $\displaystyle\int \frac{x}{\sqrt{(x+1)}} \, dx; \quad u^2 = x + 1$

11 Use the substitution $u = x - 1$ to evaluate $\displaystyle\int_2^5 \frac{x}{\sqrt{(x-1)}} \, dx$.

12 Show that $\displaystyle\int_0^1 \frac{x}{\sqrt{(1+x)}} \, dx = \tfrac{2}{3}(2 - \sqrt{2})$.

13 Use the substitution $u = \sin x$ to evaluate $\displaystyle\int_{\frac{\pi}{6}}^{\frac{\pi}{2}} \cos 2x \cos x \, dx$.

14 By using the substitution $u = \sin x$, evaluate

(a) $\displaystyle\int_0^{\frac{\pi}{2}} e^{\sin x} \cos x \, dx$

(b) $\displaystyle\int_0^{\frac{\pi}{3}} \sin 2x \sin x \, dx$

(c) $\displaystyle\int_0^{\frac{\pi}{2}} \frac{\cos x}{4 + \sin x} \, dx$

15 Evaluate (a) $\int_0^{\frac{\pi}{4}} \tan x \, \mathrm{d}x$ (b) $\int_{\frac{\pi}{4}}^{\frac{\pi}{2}} \cot x \, \mathrm{d}x$ and interpret your

answers geometrically.

4.4 Integration by parts

Remember the product formula for differentiation:

$$\frac{\mathrm{d}}{\mathrm{d}x}(uv) = v\frac{\mathrm{d}u}{\mathrm{d}x} + u\frac{\mathrm{d}v}{\mathrm{d}x}$$

You can rewrite it as:

$$v\frac{\mathrm{d}u}{\mathrm{d}x} = \frac{\mathrm{d}}{\mathrm{d}x}(uv) - u\frac{\mathrm{d}v}{\mathrm{d}x}$$

By integrating this equation, you obtain

■ $\int v\frac{\mathrm{d}u}{\mathrm{d}x}\,\mathrm{d}x = uv - \int u\frac{\mathrm{d}v}{\mathrm{d}x}\,\mathrm{d}x$

This is the formula to use when you need to integrate the product of two functions of x, v and $\frac{\mathrm{d}u}{\mathrm{d}x}$. The formula can be applied as long as you are able to differentiate v to get $\frac{\mathrm{d}v}{\mathrm{d}x}$ and you are able to integrate $\frac{\mathrm{d}u}{\mathrm{d}x}$ to get u. Even when you *can* do this, the result may not be helpful. It is regarded as helpful when $\int u\frac{\mathrm{d}v}{\mathrm{d}x}\,\mathrm{d}x$ is a less complicated expression than $\int v\frac{\mathrm{d}u}{\mathrm{d}x}\,\mathrm{d}x$. The following examples illustrate this process, which is called **integration by parts**.

Example 11

Find $\int xe^x \, \mathrm{d}x$

Take $v = x$, then $\frac{\mathrm{d}v}{\mathrm{d}x} = 1$

Take $\frac{\mathrm{d}u}{\mathrm{d}x} = e^x$, then $u = e^x$

Using the formula

$$\int v\frac{\mathrm{d}u}{\mathrm{d}x}\,\mathrm{d}x = uv - \int u\frac{\mathrm{d}v}{\mathrm{d}x}\,\mathrm{d}x$$

you have: $\int xe^x \, \mathrm{d}x = xe^x - \int e^x \cdot 1 \, \mathrm{d}x$

$$= xe^x - e^x + C$$

Example 12

Find $\displaystyle\int x \cos x \,dx$

Here you take $v = x$, then $\dfrac{dv}{dx} = 1$ and $\dfrac{du}{dx} = \cos x$, then $u = \sin x$

Applying
$$\int v \frac{du}{dx}\,dx = uv - \int u \frac{dv}{dx}\,dx$$

you get:
$$\int x \cos x \,dx = x \sin x - \int \sin x \cdot 1 \,dx$$
$$= x \sin x + \cos x + C$$

Example 13

Find $\displaystyle\int \ln x \,dx,\ x > 0$

In this case we write $v = \ln x$, then $\dfrac{dv}{dx} = \dfrac{1}{x}$

and $\dfrac{du}{dx} = 1$, then $u = x$

Using
$$\int v \frac{du}{dx}\,dx = uv - \int u \frac{dv}{dx}\,dx$$

you get:
$$\int \ln x \,dx = x \ln x - \int x \frac{1}{x}\,dx = x \ln x - x + C$$

In some cases, the process of integration by parts needs to be repeated.

Example 14

Find $\displaystyle\int x^2 \,e^{2x}\,dx$

Take $v = x^2$, then $\dfrac{dv}{dx} = 2x$ and $\dfrac{du}{dx} = e^{2x}$, then $u = \tfrac{1}{2}e^{2x}$

Applying
$$\int v \frac{du}{dx}\,dx = uv - \int u \frac{dv}{dx}\,dx$$

you get:
$$\int x^2 \,e^{2x}\,dx = \tfrac{1}{2}x^2\,e^{2x} - \int (\tfrac{1}{2}e^{2x})(2x)\,dx$$
$$= \tfrac{1}{2}x^2\,e^{2x} - \int x\,e^{2x}\,dx \qquad (1)$$

Now you need to integrate by parts again to find $\int xe^{2x}\,dx$.

Take $v = x$, then $\dfrac{dv}{dx} = 1$

and $\dfrac{du}{dx} = e^{2x}$, then $u = \frac{1}{2}e^{2x}$

Applying the 'parts' formula gives

$$\int xe^{2x}\,dx = \frac{1}{2}xe^{2x} - \int \frac{1}{2}e^{2x}\cdot 1\,dx$$
$$= \frac{1}{2}xe^{2x} - \frac{1}{4}e^{2x} \qquad\qquad (2)$$

By combining the results (1) and (2) you get

$$\int x^2 e^{2x}\,dx = \frac{1}{2}x^2 e^{2x} - \frac{1}{2}xe^{2x} + \frac{1}{4}e^{2x} + C$$
$$= \frac{1}{4}e^{2x}(2x^2 - 2x + 1) + C$$

Notice that we leave out the constant of integration until the final line.

It is easy to start applying integration by parts, and then find that the process is getting *more* complicated instead of simpler. Suppose that for $\int x^2 e^{2x}\,dx$, you took

$v = e^{2x}$ giving $\dfrac{dv}{dx} = 2e^{2x}$

and $\dfrac{du}{dx} = x^2$ giving $u = \frac{1}{3}x^3$

Then, by applying 'parts' you get

$$\int x^2 e^{2x}\,dx = \frac{1}{3}x^3 e^{2x} - \int \frac{2}{3}x^3 e^{2x}\,dx$$

which is *more* complicated than $\int x^2 e^{2x}\,dx$. In this case, you have started to work in the wrong direction. You would have to start again with $v = x^2$ and $\dfrac{du}{dx} = e^{2x}$, as shown in Example 14.

Exercise 4D

Use integration by parts to find:

1 $\displaystyle\int xe^{-x}\,dx$ **2** $\displaystyle\int xe^{3x}\,dx$ **3** $\displaystyle\int x\sin x\,dx$

4 $\displaystyle\int x\ln x\,dx$ **5** $\displaystyle\int \ln(x-1)\,dx$ **6** $\displaystyle\int x\cos 3x\,dx$

7 $\displaystyle\int x(x-1)^4\,\mathrm{d}x$ **8** $\displaystyle\int x\sqrt{(x-1)}\,\mathrm{d}x$ **9** $\displaystyle\int x^2\mathrm{e}^x\,\mathrm{d}x$

10 $\displaystyle\int x^2\cos x\,\mathrm{d}x$ **11** $\displaystyle\int x^2\mathrm{e}^{-x}\,\mathrm{d}x$ **12** $\displaystyle\int x^3\ln x\,\mathrm{d}x$

Evaluate each of the following definite integrals:

13 $\displaystyle\int_0^{\pi} x\sin x\,\mathrm{d}x$ **14** $\displaystyle\int_0^{\frac{\pi}{2}} x\cos\tfrac{1}{2}x\,\mathrm{d}x$

15 $\displaystyle\int_1^{\mathrm{e}} x^2\ln x\,\mathrm{d}x$ **16** $\displaystyle\int_0^1 x(x-1)^3\,\mathrm{d}x$

17 $\displaystyle\int_0^2 (x-1)(x+1)^3\,\mathrm{d}x$ **18** $\displaystyle\int_1^{\mathrm{e}} \frac{\ln x}{x^4}\,\mathrm{d}x$

19 $\displaystyle\int_1^{\mathrm{e}} (\ln x)^2\,\mathrm{d}x$ **20** $\displaystyle\int_0^{\frac{\pi}{2}} \mathrm{e}^x\sin x\,\mathrm{d}x$

4.5 A systematic approach to integration

At this stage, you should review the various methods of integration that you have met in this chapter and in Books P1 and P2.

■ **The standard forms given at the end of this chapter in the summary should all be memorised.**

You should be able to recognise expressions that can be integrated using standard forms at once.

Example 15

(a) $\displaystyle\int (4x-1)^5\,\mathrm{d}x = \frac{(4x-1)^6}{4\times 6} + C$

$$= \tfrac{1}{24}(4x-1)^6 + C$$

(b) $\displaystyle\int \mathrm{e}^{-4x}\,\mathrm{d}x = -\tfrac{1}{4}\mathrm{e}^{-4x} + C$

(c) $\displaystyle\int \sin\tfrac{1}{3}x\,\mathrm{d}x = -3\cos\tfrac{1}{3}x + C$

■ **You should be able to convert expressions to equivalent forms that *can* be integrated by using trigonometric identities and partial fractions.**

Example 16

(a) $\int \tan^2 3x \, dx = \int (\sec^2 3x - 1) \, dx$

$$= \tfrac{1}{3} \tan 3x - x + C$$

(b) $\int \dfrac{1}{x(x+1)} \, dx = \int \left(\dfrac{1}{x} - \dfrac{1}{x+1} \right) dx$

$$= \ln |x| - \ln |x+1| + C$$

■ **You should recognise an expression that is given as a function and its derivative.**

Example 17

(a) $\int x e^{x^2} \, dx = \tfrac{1}{2} e^{x^2} + C$

$\left(\begin{array}{l} \text{function of } x^2 \\ \text{derivative } 2x \end{array} \right)$ OR   the substitution $u = x^2$ could be used

(b) $\int \dfrac{x}{(x^2 + 1)^{\frac{1}{2}}} \, dx = (x^2 + 1)^{\frac{1}{2}} + C$

$\left(\begin{array}{l} \text{function of } x^2 \\ \text{derivative } 2x \end{array} \right)$ OR   the substitution $u = x^2 + 1$ could be used

(c) $\int \tan^3 2x \sec^2 2x \, dx = \tfrac{1}{8} \tan^4 2x + C$

$\left(\begin{array}{l} \text{function of } \tan 2x \\ \text{derivative } 2 \sec^2 2x \end{array} \right)$ OR   the substitution $u = \tan 2x$ could be used

■ **You should recognise when you need to use integration by parts.**

Look back at the examples given in section 4.4.

In your work you will need to identify which approach is required and choose your method accordingly. In some cases you may be given a hint about the method or the substitution to employ. Practice is the key to success and the following exercises are typical of what is expected by Edexcel examiners.

Exercise 4E

Integrate with respect to x:

1 $(4x + 5)^{\frac{1}{2}}$

2 $\dfrac{1}{4x + 5}$

3 $\left(1 - \dfrac{1}{x} \right)^2$

4 $\cos x \sin x$

5 $\tan 3x$

6 $x \sin 3x$

7 $\dfrac{1 + x}{x^{\frac{1}{2}}}$

8 $\dfrac{x}{1 + x}$

9 $\sin x \cos^4 x$

10 $3 \ln x$

11 $\dfrac{x+2}{x(x-1)}$

12 $\dfrac{\sec^2 x}{(1+\tan x)^3}$

13 $\sin^2 2x$

14 $\dfrac{x^2}{x-2}$

15 $(\sin x + 2\cos x)^2$

16 $x^2 e^{\frac{1}{2}x}$

17 $\dfrac{1}{x^2-4}$

18 $\dfrac{x}{9x^2+1}$

19 $(1-x^{-2})^2$

20 $(2-3x)^{-2}$

21 $(4-5x)^{-1}$

22 $\cot 3x$

23 $\operatorname{cosec} 2x \cot 2x$

24 $\cot^2 3x$

25 $x \cos 5x$

26 $\dfrac{x}{(x-1)^{\frac{1}{2}}}$

27 $x^2 e^{-x}$

28 $\cos 2x \sin x$

29 $\sin 2x \cos x$

30 $\tan 2x \sec 2x$

31 $\dfrac{(x+1)^2}{x^2+1}$

32 $\dfrac{2}{(x-2)(x-4)}$

33 $\dfrac{1}{x^2(x-1)}$

34 $\operatorname{cosec}^2 2x + 1$

35 $\dfrac{x+4}{x-4}$

36 $\dfrac{1}{x(x^2-1)}$

37 $\dfrac{x^2}{x^3+1}$

38 $(e^x + x)^2$

39 $x^3 \ln x$

40 $x^3 e^{x^2}$

41 Use the identity $\cos^2 x + \sin^2 x \equiv 1$ and the substitution $\cos x = u$ to find $\displaystyle\int \sin^3 x \,\mathrm{d}x$.

42 Find $\displaystyle\int \cos^3 x \,\mathrm{d}x$ and $\displaystyle\int \sin^5 x \,\mathrm{d}x$.

43 Use the identity $\sec^2 x \equiv \tan^2 x + 1$ and the substitution $\tan x = u$ to find $\displaystyle\int \tan^4 x \,\mathrm{d}x$.

44 Find (a) $\displaystyle\int \sec^4 x \,\mathrm{d}x$ (b) $\displaystyle\int \cot^4 x \,\mathrm{d}x$.

45 Use the identity $\sin(A+B) + \sin(A-B) \equiv 2\sin A \cos B$ to find

(a) $\displaystyle\int 2\sin 6x \cos 4x \,\mathrm{d}x$

(b) $\displaystyle\int \sin x \cos\tfrac{1}{2}x \,\mathrm{d}x$

46 Evaluate $\displaystyle\int_0^1 \dfrac{x+9}{(x+2)(3-2x)} \,\mathrm{d}x$.

47 Use the substitution $x = 3\sin t$ to show that

$$\int_0^3 x^2(9-x^2)^{\frac{1}{2}} \,\mathrm{d}x = \tfrac{81}{16}\pi$$

48 Evaluate $\displaystyle\int_{\frac{\pi}{6}}^{\frac{\pi}{3}} \sec^3 x \tan x \,\mathrm{d}x$.   **49** Evaluate $\displaystyle\int_{3}^{4} \frac{2x+4}{(x-2)(x^2+4)} \,\mathrm{d}x$

50 Evaluate $\displaystyle\int_{1}^{2} \frac{x}{(1+x^2)} \,\mathrm{d}x$.

4.6 Further examples on areas of regions and volumes of solids of revolution

In Book P2 chapter 7 you were given these formulae:

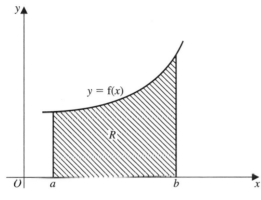

$$\text{Area of } R = \int_{a}^{b} \mathrm{f}(x) \,\mathrm{d}x$$

Volume generated when R is rotated completely about the x-axis is

$$\pi \int_{a}^{b} [\mathrm{f}(x)]^2 \,\mathrm{d}x$$

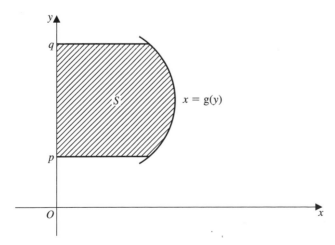

$$\text{Area of } S = \int_{p}^{q} \mathrm{g}(y) \,\mathrm{d}y$$

Volume generated when S is rotated completely about the y-axis is

$$\pi \int_p^q [g(y)]^2 dy$$

Now that you have learned in this chapter how to evaluate many more integrals, you will be able to find the exact areas and volumes resulting from many more curves.

Example 18

The region R is bounded by the curve with equation $y = \cos^2 2x$, the x-axis and the lines $x = 0$ and $x = \dfrac{\pi}{6}$. Find the area of R.

First draw a sketch, showing R:

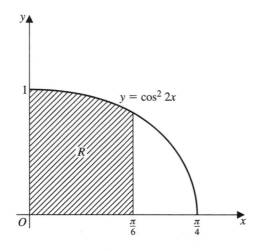

$$\text{Area of } R = \int_0^{\frac{\pi}{6}} \cos^2 2x \, dx$$

Use the identity $\cos 2A \equiv 2\cos^2 A - 1$, putting $A = 2x$:

$$\cos 4x \equiv 2\cos^2 2x - 1$$

$$\Rightarrow \qquad \cos^2 2x \equiv \frac{1 + \cos 4x}{2}$$

So: $\qquad \text{area of } R = \displaystyle\int_0^{\frac{\pi}{6}} (\tfrac{1}{2} + \tfrac{1}{2}\cos 4x) \, dx$

$$\text{Area of } R = \left[\tfrac{1}{2}x + \tfrac{1}{8}\sin 4x \right]_0^{\frac{\pi}{6}}$$

$$= \frac{\pi}{12} + \tfrac{1}{8}\sin \frac{2\pi}{3} - (0 + 0)$$

$$= \frac{\pi}{12} + \frac{\sqrt{3}}{16} \quad \left(\text{since } \sin \frac{2\pi}{3} = \frac{\sqrt{3}}{2} \right)$$

Example 19

The region R shown is bounded by the curve with equation $y = \ln x$, the y-axis and the lines $y = 2$ and $y = 5$.

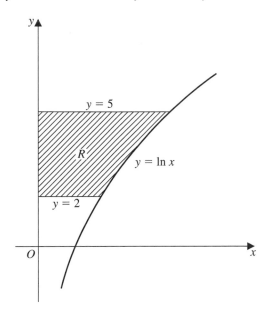

Find
(a) the area of R (b) the volume generated when R is rotated through 2π about the y-axis, giving your answers to 3 significant figures.

(a) Rewrite the equation $y = \ln x$ in the form $x = e^y$ because to find the area of R you need to evaluate

$$\int_2^5 x\,\mathrm{d}y = \int_2^5 e^y\,\mathrm{d}y$$

$$\text{Area of } R = \left[e^y \right]_2^5 = e^5 - e^2 \approx 141 \text{ units}^2$$

(b) The volume generated when R is rotated through 2π about the y-axis is

$$\int_2^5 \pi x^2\,\mathrm{d}y = \pi \int_2^5 e^{2y}\,\mathrm{d}y$$

$$= \pi \left[\tfrac{1}{2} e^{2y} \right]_2^5$$

$$= \frac{\pi}{2}\left(e^{10} - e^4 \right) \approx 34\,500 \text{ units}^3$$

Example 20

The normal to the curve with equation $y = e^x$ at the point $B(1, e)$ meets the x-axis at the point C. The finite region bounded by the curve, the line BC, the y-axis and the x-axis is rotated through 2π radians about the x-axis. Find the volume of revolution so generated.

For $y = e^x$,

$$\frac{dy}{dx} = e^x = e \text{ at } x = 1$$

So the gradient of the normal to $y = e^x$ at $B(1, e)$ is $-\dfrac{1}{e}$.

An equation of the normal to $y = e^x$ at B is

$$y - e = -\frac{1}{e}(x - 1)$$

The normal meets the x-axis at $y = 0$.

So:
$$0 - e = -\frac{1}{e}(x - 1)$$

$$x - 1 = e^2$$

and:
$$x = e^2 + 1$$

The point C is $(e^2 + 1, 0)$.

The region to be rotated looks like this:

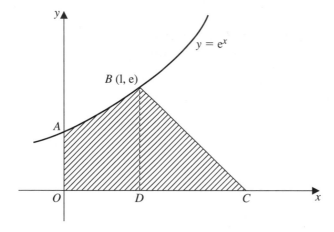

Part of the region lies under the curve from A to B and the rest lies under the line BC. The point D on the x-axis has the same x-coordinate as B; that is, it is at $(1, 0)$. When $\triangle BDC$ is rotated about the x-axis it forms a cone of base radius $BD = e$ and height $DC = e^2$ because

$$DC = OC - OD = (1 + e^2) - 1 = e^2$$

The volume of a cone is $\frac{1}{3}$(base area) $\times$ height and the volume of the cone formed by the complete rotation of $\triangle DBC$ about the x-axis is

$$\tfrac{1}{3}\pi \times BD^2 \times DC = \tfrac{1}{3}\pi \times e^2 \times e^2 = \tfrac{1}{3}\pi e^4$$

The volume generated when the region $ABDO$ is rotated completely about the x-axis is:

$$\int_0^1 \pi(e^x)^2 \, dx = \int_0^1 \pi e^{2x} \, dx$$

$$= \tfrac{1}{2}\pi \left[e^{2x} \right]_0^1$$

$$= \tfrac{1}{2}\pi(e^2 - 1)$$

Total volume generated $= \tfrac{1}{3}\pi e^4 + \tfrac{1}{2}\pi e^2 - \tfrac{1}{2}\pi$

$$= \tfrac{1}{6}\pi(2e^4 + 3e^2 - 3)$$

Example 21

The curve given parametrically by

$$x = \tan t, \ y = \sin t, \ 0 \leqslant t < \tfrac{\pi}{2}$$

is sketched in the diagram.

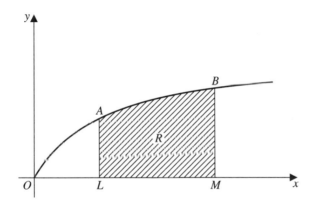

At the points A and B on the curve $t = \dfrac{\pi}{6}$ and $\dfrac{\pi}{3}$ respectively. The lines AL and BM are drawn parallel to the y-axis, as shown. Determine the area of the finite region R, shown shaded, and find the volume generated when R is rotated completely about the x-axis.

Area of $R = \displaystyle\int_{t=\frac{\pi}{6}}^{t=\frac{\pi}{3}} y \, dx$ and we need to express both y and dx in terms of t.

We know that $y = \sin t$ and that, by differentiation, $\dfrac{dx}{dt} = \sec^2 t$.

Also, the integral $\displaystyle\int y \, dx$ can be written as

$$\int y \cdot \frac{dx}{dt} \cdot dt$$

by using the chain rule.

So:

$$\text{Area of } R = \int_{\frac{\pi}{6}}^{\frac{\pi}{3}} \sin t \sec^2 t \, dt$$

$$= \int_{\frac{\pi}{6}}^{\frac{\pi}{3}} \frac{\sin t}{\cos^2 t} \, dt$$

This integral is a function of $\cos t$ and its derivative $-\sin t$.

Putting $u = \cos t$, $\dfrac{du}{dt} = -\sin t$

and when $t = \dfrac{\pi}{3}, u = \dfrac{1}{2}$; $t = \dfrac{\pi}{6}$, $u = \dfrac{\sqrt{3}}{2}$

So:

$$\text{Area of } R = \int_{\frac{\sqrt{3}}{2}}^{\frac{1}{2}} \frac{1}{u^2} (-du)$$

$$= -\int_{\frac{\sqrt{3}}{2}}^{\frac{1}{2}} u^{-2} \, du$$

$$= \left[u^{-1} \right]_{\frac{\sqrt{3}}{2}}^{\frac{1}{2}} = 2 - \frac{2}{\sqrt{3}}$$

$$\text{Volume generated} = \int_{t=\frac{\pi}{6}}^{t=\frac{\pi}{3}} \pi y^2 \, dx$$

$$= \pi \int_{t=\frac{\pi}{6}}^{t=\frac{\pi}{3}} y^2 \frac{dx}{dt} \cdot dt$$

$$= \pi \int_{\frac{\pi}{6}}^{\frac{\pi}{3}} \sin^2 t \sec^2 t \, dt$$

$$= \pi \int_{\frac{\pi}{6}}^{\frac{\pi}{3}} \frac{\sin^2 t}{\cos^2 t} \, dt = \pi \int_{\frac{\pi}{6}}^{\frac{\pi}{3}} \tan^2 t \, dt$$

You can use the identity $\sec^2 t \equiv 1 + \tan^2 t$

so: $$\tan^2 t \equiv \sec^2 t - 1$$

Thus:

$$\text{Volume generated} = \pi \int_{\frac{\pi}{6}}^{\frac{\pi}{3}} (\sec^2 t - 1) \, dt$$

$$= \pi \left[\tan t - t \right]_{\frac{\pi}{6}}^{\frac{\pi}{3}}$$

$$= \pi \left[\tan \frac{\pi}{3} - \frac{\pi}{3} - \left(\tan \frac{\pi}{6} - \frac{\pi}{6} \right) \right]$$

$$= \pi \left[\sqrt{3} - \frac{\pi}{3} - \frac{1}{\sqrt{3}} + \frac{\pi}{6} \right] = \pi \left[\frac{2}{\sqrt{3}} - \frac{\pi}{6} \right]$$

Exercise 4F

In questions 1–10, find the area of the region bounded by the curve with equation $y = f(x)$, the x-axis and the lines $x = a$ and $x = b$.

1 $f(x) = \cos x$, $a = \dfrac{\pi}{6}$, $b = \dfrac{\pi}{3}$

2 $f(x) = \sec^2 x$, $a = 0$, $b = \dfrac{\pi}{4}$

3 $f(x) = xe^x$, $a = 1$, $b = 3$

4 $f(x) = \ln x$, $a = 2$, $b = 5$

5 $f(x) = \sin^2 x$, $a = \dfrac{\pi}{12}$, $b = \dfrac{5\pi}{12}$

6 $f(x) = \tan^2 x$, $a = \dfrac{\pi}{8}$, $b = \dfrac{3\pi}{8}$

7 $f(x) = x \cos 2x$, $a = \dfrac{\pi}{6}$, $b = \dfrac{\pi}{5}$

8 $f(x) = xe^{x^2}$, $a = -2$, $b = -1$

9 $f(x) = \dfrac{\sin x}{2 + \cos x}$, $a = \dfrac{\pi}{2}$, $b = \dfrac{2\pi}{3}$

10 $f(x) = \tan^2 x \sec^2 x$, $a = \dfrac{\pi}{6}$, $b = \dfrac{\pi}{3}$

In questions 11–20, the finite region R is bounded by the curve with equation $y = f(x)$, the x-axis and the lines $x = a$ and $x = b$. Find the volume generated when R is rotated completely about the x-axis.

11 $f(x) = x^{\frac{1}{2}}$, $a = 0$, $b = 4$

12 $f(x) = 2x^{\frac{1}{4}}$, $a = 1$, $b = 16$

13 $f(x) = \sin x$, $a = 0$, $b = \pi$

14 $f(x) = x^{-\frac{1}{2}}$, $a = 2$, $b = 5$

15 $f(x) = x^{\frac{1}{2}}e^x$, $a = 1$, $b = 2$

16 $f(x) = x^2 - 4$, $a = 3$, $b = 5$

17 $f(x) = \cot x$, $a = \frac{\pi}{4}$, $b = \frac{\pi}{2}$

18 $f(x) = \ln x$, $a = 1$, $b = 3$

19 $f(x) = \dfrac{x+1}{x}$, $a = 1$, $b = 4$

20 $f(x) = x\sqrt{(4 - x^2)}$, $a = 0$, $b = 2$

21 The finite region bounded by the curve with equation $y = \tan \frac{1}{2}x$, the line $x = \frac{\pi}{2}$ and the coordinate axes is rotated through 2π radians about the x-axis. Show that the volume generated is $\frac{\pi}{2}(4 - \pi)$.

22 By rotating the semi-circle for which $y \geqslant 0$ from the circle with equation $x^2 + y^2 = a^2$ completely about the x-axis, show that the volume of a sphere, of radius a, is $\frac{4}{3}\pi a^3$.

23 A triangular region is bounded by the line with equation $y = \dfrac{r}{h}x$, where r and h are positive constants, the line $x = r$ and the x-axis. By considering the complete rotation of this triangle about the x-axis, show that the volume of a cone of height h and base radius r is $\frac{1}{3}\pi r^2 h$.

24 Find the area of the region bounded by the curve with equation $y^3 = x$, the lines $y = 2$, $y = 4$ and $x = 0$. This region is rotated completely about the y-axis. Find the volume generated.

25 The region R shown in the figure is bounded by the curve with equation $y = \cos x - \sin x$ and the coordinate axes.

(a) Find the area of R.

The region R is rotated completely about the x-axis.

(b) Find the volume generated.

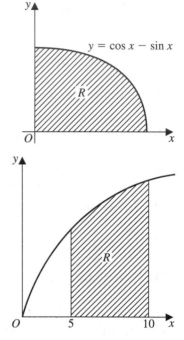

26 The diagram shows a sketch of the curve with equation

$$x = t^2 + 1, \ y = t$$

where t is a parameter for $t > 0$.

The shaded region R is bounded by the curve, the x-axis and the lines $x = 5$ and $x = 10$.

(a) Find the area of R.

(b) Find the volume generated when R is rotated through 2π about the x-axis.

27 Find the area of the finite region R bounded by the curve with parametric equations $x = 1 + t$, $y = 4 - t^2$, the lines $x = 3$ and $x = 6$ and the x-axis.

Find also the volume generated when R is rotated completely about the x-axis.

28 Find the area of the finite region R bounded by the curve with parametric equations $x = 4t$, $y = \dfrac{4}{t}$, the x-axis and the lines $x = 3$ and $x = 16$.

The region R is rotated through 2π about the x-axis. Find the volume of the solid generated.

29 An ellipse is given by

$$x = 3\cos t, \ y = 2\sin t, \ 0 \leqslant t < 2\pi$$

(a) Find the area of the finite region bounded by the ellipse and the positive x- and y-axes.

(b) This region is rotated completely about the y-axis to form a solid of revolution. Find the volume of this solid.

4.7 Exponential growth and decay

Forming and solving simple differential equations

Any equation involving derivatives of one variable with respect to another variable is called a **differential equation**. Simple examples are

$$\frac{dy}{dx} = x^2 - 5 \quad \text{and} \quad \frac{dy}{dx} = \sin y \cos x$$

These equations are called **first order differential equations** because the highest derivative each contains is the first derivative of y with respect to x, that is $\frac{dy}{dx}$.

A **second order differential equation** might look like this:

$$\frac{d^2y}{dx^2} + x^2 \frac{dy}{dx} = y$$

In Book P3, you will only be concerned with first order differential equations of a special simple type. In Books P4, P5 and P6, you will meet other differential equations.

In many practical situations, the rate at which one variable is changing with respect to another is expressed by a **physical law**. A differential equation can be set up from the experimental data and possible solutions of this equation are found using integration. One of the most important and widely occurring relationships between variables is called the **law of natural growth or decay** or **exponential growth and decay**. Here are a few examples.

(a) Population growth

People such as scientists, sociologists and town planners are often more concerned with the *rate* at which a particular quantity is growing than with its current size. The Director of Education is more concerned with the rate at which the school population is increasing or decreasing than with what the population is now, because he has to plan for the future and ensure that there are enough (and not too many) school places available to meet demand each year. The scientist may need to know the rate at which a colony of bacteria is growing rather than how many of the bacteria exist at this moment.

One thing that each of these populations has in common is that their rate of increase is proportional to the size of the population at any time. Now in Book P1 you were shown that $\frac{dy}{dx}$ represents

130 Integration

the rate of change of y with respect to x. So if the size of a population at a given time t is P then the rate of increase of the population is a short way of saying 'the rate of increase of the population as time goes on'. That is, it is the rate of change of P with respect to t, i.e. $\dfrac{dP}{dt}$. Now if this rate of increase is proportional to the size of the population at any given time, then $\dfrac{dP}{dt} = kP$, where k is a constant. If you take the equation

$$\frac{dP}{dt} = kP$$

and divide both sides by P you get

$$\frac{1}{P}\frac{dP}{dt} = k$$

If you now integrate both sides with respect to t, you get

$$\int \frac{1}{P}\frac{dP}{dt}\,dt = \int k\,dt$$

or:

$$\int \frac{1}{P}\,dP = \int k\,dt$$

i.e.

$$\ln P = kt + C$$

where C is a constant.

Now

$$\ln P = \log_e P$$

So:

$$\log_e P = kt + C$$

and:

$$P = e^{kt+C} = e^{kt}\cdot e^{C}$$

Since both e and C are constants, it follows that e^C is a constant. Call this constant A.

Then:

$$P = Ae^{kt}$$

This demonstrates that the population grows **exponentially**; that is, P is a function of e^t.

(b) Radioactive decay

In a mass of radioactive material, where the atoms are disintegrating spontaneously, the average rate of disintegration is proportional to the number of atoms present. At time t, there are N atoms present and this situation is described by the differential equation

$$\frac{dN}{dt} = -kN$$

where k is a constant. The minus sign indicates that the rate of change is *decreasing*.

(c) Newton's law of cooling

The rate of change of the temperature of a cooling body is proportional to the excess temperature over the surroundings. If the excess temperature is θ at time t, then this situation is described by the differential equation

$$\frac{d\theta}{dt} = -k\theta$$

where k is a positive constant.

(d) Chemical reactions

Some chemical reactions follow a law which states that the rate of change of the reacting substance is proportional to its concentration. If the concentration is C at time t, then this situation is described by the differential equation

$$\frac{dC}{dt} = -kC$$

where k is a positive constant.

As you can see one common differential equation can be used to model the rates of change of several quite different variables with respect to time. These variables come from a whole range of different physical situations.

In your exam, you may be given the description of a law in words and you will be required to form a differential equation from the description. You will not require any special knowledge to do this, except to recognise that the rate of change of a variable y with respect to x is $\frac{dy}{dx}$ and that if A is proportional to B, then $A = kB$ where k is a constant.

Example 22

The length y cm of a leaf during the period of its growth is proportional to the amount of water it contains. During this period the leaf retains a similar shape; that is, the ratio of its length to its width remains constant. The leaf absorbs water from its 'parent' plant at a rate proportional to y and it loses water by evaporation at a rate proportional to the area of the leaf at the time when its length is y cm. Form a differential equation to describe the growth of the leaf.

Assume that the leaf has length y cm at time t days after it was first observed.

The rate at which the leaf is receiving water is $k_1 y$ where k_1 is a positive constant.

The area of the leaf at time t days is proportional to y^2, since it maintains its shape.

So the leaf is losing water at a rate of $k_2 y^2$, where k_2 is another positive constant.

The rate of growth of the leaf is given by $\dfrac{dy}{dt}$, the rate of change of its length.

$$\frac{dy}{dt} = k_1 y - k_2 y^2$$

is a differential equation describing the growth of the leaf.

In this chapter you learn how to solve differential equations of the first order in which the variables are separable. This type of differential equation is of the form

$$\frac{dy}{dx} = f(x)g(y)$$

First, let's look at two simpler cases.

(i) Suppose that $g(y) = 1$, then:

$$\frac{dy}{dx} = f(x)$$

By direct integration:

$$y = \int f(x)\, dx + C$$

where C is a constant.

$y = \int f(x)\, dx + C$ is called **the general solution of the differential equation** $\dfrac{dy}{dx} = f(x)$, once the integration is completed.

Example 23

Solve the differential equation $\dfrac{dy}{dx} = \ln x$, $x > 0$, given that $y = 2$ at $x = 1$.

We have $y = \int \ln x\, dx + C$, where C is a constant. From example 13 of this chapter,

$$\int \ln x\, dx = x \ln x - x$$

and therefore $y = x \ln x - x + C$ is the general solution of the differential equation.

At $x = 1$, $y = 2 \Rightarrow 2 = 1 \cdot \ln 1 - 1 + C$

So: $\qquad\qquad\qquad\qquad C = 3$

The solution of the differential equation is:

$$y = x \ln x - x + 3$$

(ii) Suppose now that $f(x) = 1$ so that the differential equation is

$$\frac{dy}{dx} = g(y)$$

$\dfrac{dy}{dx} = \dfrac{1}{\frac{dx}{dy}}$, so:

$$\frac{dx}{dy} = \frac{1}{g(y)}$$

and

$$x = \int \frac{1}{g(y)} \, dy + C$$

where C is a constant.

That is, provided you can integrate $\dfrac{1}{g(y)}$ with respect to y, you have found the general solution.

Example 24

Given that $y > -\frac{1}{2}$, solve the differential equation $\dfrac{dy}{dx} = 2y + 1$.

Express the general solution in the form $y = f(x)$.

$$\frac{dy}{dx} = 2y + 1 \Rightarrow \frac{dx}{dy} = \frac{1}{2y + 1}$$

Integrating: $x = \displaystyle\int \frac{1}{2y + 1} \, dy = \frac{1}{2}\ln|2y + 1| + C$

The general solution is then

$$x = \tfrac{1}{2}\ln|2y + 1| + C$$

You now want y in terms of x.

Rearranging: $\ln|2y + 1| = 2(x - C)$

$\Rightarrow$ $2y + 1 = e^{2(x - C)}$

and $y = \tfrac{1}{2}(e^{2x - 2C} - 1)$ is the form required.

The differential equation

$$\frac{dy}{dx} = f(x)g(y)$$

can be written as

$$\frac{1}{g(y)} \frac{dy}{dx} = f(x)$$

Integrating with respect to x gives:

$$\int \frac{1}{g(y)} \frac{dy}{dx} \cdot dx = \int f(x)\, dx + C$$

- **That is,** $$\int \frac{1}{g(y)}\, dy = \int f(x)\, dx + C$$

   is the general solution, provided that $\dfrac{1}{g(y)}$ **can be integrated with respect to y and $f(x)$ can be integrated with respect to x.**

Differential equations of the type $\dfrac{dy}{dx} = f(x)g(y)$ are known as **first order separable** because, as you have seen, they can be solved by *separating* the variables and integrating.

Example 25

Express y in terms of x, given that $\dfrac{dy}{dx} = (y+2)(2x+1)$ and that $y = 2$ at $x = 0$.

Rewrite the differential equation as:

$$\frac{1}{y+2} \frac{dy}{dx} = 2x + 1$$

Integrating gives $\qquad \ln|y+2| = x^2 + x + C$

as the general solution.

At $x = 0$, $y = 2 \Rightarrow \ln 4 = 0 + 0 + C$

So: $\qquad\qquad\qquad\qquad C = \ln 4$

and: $\qquad\qquad \ln|y+2| - \ln 4 = x^2 + x$

$$\ln\left|\frac{y+2}{4}\right| = x^2 + x$$

$$\frac{y+2}{4} = e^{x^2 + x}$$

And so $\qquad\qquad y = 4e^{x^2 + x} - 2$

is the required solution.

Example 26

(a) Newton's law of cooling states that the rate at which the temperature of a preheated body decreases is proportional to the difference between the temperature of the body and that of the surroundings. Given that $\theta\,°C$ is the excess of the temperature of the body over that of the surroundings at time t minutes after the start, show that the relationship between θ and t is of the form $\theta = Ae^{-kt}$ where A and k are constants.

(b) A bowl of water whose temperature is $\phi\,°C$, at time t minutes is placed in a room where the temperature remains constant at $15\,°C$. Given that $\phi = 100$ at $t = 0$, and $\phi = 40$ at $t = 30$, find the temperature of the water at time (i) $t = 10$ (ii) $t = 45$. Find also the time when (iii) $\phi = 70$ (iv) $\phi = 53$.

(a) If the temperature of the water at time t minutes is $\phi\,°C$ then the rate of change of the temperature of the water is $\dfrac{d\phi}{dt}$

But $\theta = \phi - M$, where $M\,°C$ is the constant temperature of the surroundings. Differentiating with respect to t gives

$$\frac{d\theta}{dt} = \frac{d\phi}{dt}$$

since M is a constant.

So the rate of change of the temperature of the water can also be written as $\dfrac{d\theta}{dt}$.

The rate at which the temperature of the water *decreases* is therefore $-\dfrac{d\theta}{dt}$. (The minus sign indicates that the temperature is decreasing rather than increasing.)

If this rate is proportional to θ then

$$-\frac{d\theta}{dt} = k\theta$$

where k is a constant.

i.e.

$$\frac{d\theta}{dt} = -k\theta$$

$$\frac{1}{\theta}\frac{d\theta}{dt} = -k$$

$$\int \frac{1}{\theta}\frac{d\theta}{dt}\, dt = \int -k\, dt$$

$$\int \frac{1}{\theta}\, d\theta = \int -k\, dt$$

So:

$$\ln|\theta| = -kt + C$$

where C is a constant.

$$\theta = e^{-kt+C} = e^{-kt}e^{C}$$

$$\theta = Ae^{-kt}, \text{ where } A = e^{C}$$

The minus sign indicates that the temperature of the water is *decreasing*. This is an example of **exponential decay**.

(b) When $t = 0$, $\phi = 100$, $M = 15$ and

$$\theta = 100 - 15 = 85$$

So:
$$85 = Ae^0 \qquad (1)$$

$$A = 85$$

When $t = 30$, $\phi = 40$ and

$$\theta = 40 - 15 = 25$$

So:
$$25 = Ae^{-30k} \qquad (2)$$

Substitute $A = 85$ in (2):

$$25 = 85e^{-30k}$$

$$e^{-30k} = \frac{25}{85}$$

$$\ln e^{-30k} = \ln \frac{25}{85}$$

$$-30k = -1.2237 \text{ (since } \ln e = 1)$$

$$k = 0.04 \text{ (1 s.f.)}$$

So:
$$\theta = 85e^{-0.04t}$$

(i) $t = 10 \Rightarrow \theta = 85e^{-0.4} \approx 57$

$$57 \approx \phi - 15$$

$$\phi \approx 72$$

(ii) $t = 45 \Rightarrow \theta = 85e^{-1.8} \approx 14$

$$14 \approx \phi - 15$$

$$\phi \approx 29$$

(iii) $\phi = 70 \Rightarrow \theta = 55$

$$55 = 85e^{-0.04t}$$

$$e^{-0.04t} = \frac{55}{85}$$

$$-0.04t = \ln \frac{55}{85}$$

$$t = -\frac{1}{0.04} \ln \frac{55}{85}$$

$$t \approx 11$$

(iv) $\phi = 53 \Rightarrow \theta = 38$

$$38 = 85e^{-0.04t}$$

$$-0.04t = \ln \frac{38}{85}$$

$$t \approx 20$$

Exercise 4G

Find the general solutions of the differential equations in questions 1–10.

1 $\dfrac{dy}{dx} = e^{2x-1}$ 　　　　　　**2** $\dfrac{dy}{dx} = e^{2y-1}$

3 $\dfrac{dy}{dx} = \cos^2 x$ 　　　　　　**4** $\dfrac{dy}{dx} = \cos^2 y$

5 $\dfrac{dy}{dx} = xy$ 　　　　　　**6** $\dfrac{dy}{dx} = e^{x+y}$

7 $\dfrac{dy}{dx} = \sec y \ln x$ 　　　　　**8** $(x+1)\dfrac{dy}{dx} = y+2$

9 $x\dfrac{dy}{dx} = y + x^2 y$ 　　　　**10** $y\dfrac{dy}{dx} + \cot x \operatorname{cosec} x = 0$

Obtain the solution that satisfies the given conditions of the differential equations in questions 11–20.

11 $\dfrac{dy}{dx} = x^2 + x, \quad y = 0 \text{ at } x = 0$

12 $\dfrac{dy}{dx} = \sin^2 x \cos x, \quad y = 0 \text{ at } x - \dfrac{\pi}{2}$

13 $\dfrac{dy}{dx} - 3y + 1, \quad y = 0 \text{ at } x = 1$

14 $\dfrac{dy}{dx} = xe^y, \quad y = 0 \text{ at } x = 1$

15 $\dfrac{dy}{dx} = \tan x \tan y, \quad y = \dfrac{\pi}{4} \text{ at } x = \dfrac{\pi}{4}$

16 $\dfrac{dy}{dx} = \sin^2 x \cos^2 y, \quad y = 0 \text{ at } x = 0$

17 $x^2\dfrac{dy}{dx} = \operatorname{cosec} y \sec y, \quad y = \dfrac{\pi}{3} \text{ at } x = 1$

18 $y\dfrac{dy}{dx} = \sec^2 x(2\tan x + 1), \ y = 3 \text{ at } x = \dfrac{\pi}{4}$

19 $y\sin y\dfrac{dy}{dx} = x\cos x, \ y = 0 \text{ at } x = \dfrac{\pi}{2}$

20 $\dfrac{dy}{dx} = (y^2 - 1)\cot x, \ y = 2 \text{ at } x = \dfrac{\pi}{4}$

21 Newton's law of cooling states that the rate of change of the temperature of a cooling liquid is proportional to the excess temperature over the room temperature. The law is given by the differential equation

$$\dfrac{d\theta}{dt} = -k\theta$$

where θ is the excess temperature at time t. At $t = 0$, $\theta = \theta_0$.
Show that

$$\theta = \theta_0 e^{-kt}$$

22 The temperature of a liquid in a room, where the temperature
is constant at $20\,°C$, was observed to be $80\,°C$ and 7 minutes
later it was $60\,°C$. Calculate, using Newton's law of cooling,
(a) the time taken for the temperature to fall from $80\,°C$ to
$40\,°C$
(b) the temperature of the liquid 10 minutes after it was $80\,°C$.

23 A lump of radioactive substance is disintegrating. At time t
days after it was first observed to have mass 10 grams, its
mass is m grams and

$$\frac{dm}{dt} = -km$$

where k is a positive constant.
Find the time, in days, for the substance to reduce to 1 gram
in mass, given that its half-life is 8 days.
(The half-life is the time in which half of any mass of the
substance will decay.)

24 Given that $x\dfrac{dy}{dx} = (1 - 2x^2)y$, where $x > 0$, and that $y = 1$ at
$x = 1$, find y in terms of x.

25 The gradient at any point (x, y), where $x > 0$, on a curve is
$\ln x$ and $y = e$ at $x = 1$. Find the equation of the curve.

SUMMARY OF KEY POINTS

1 Standard results (to be memorised):

$$\int \sin x \, dx = -\cos x + C$$

$$\int \sin(ax + b) \, dx = -\frac{1}{a}\cos(ax + b) + C$$

$$\int \cos x \, dx = \sin x + C$$

$$\int \cos(ax + b) \, dx = \frac{1}{a}\sin(ax + b) + C$$

$$\int (ax + b)^n \, dx = \frac{1}{a(n+1)}(ax + b)^{n+1} + C, n \neq -1$$

$$\int \frac{1}{ax + b} \, dx = \frac{1}{a}\ln|ax + b| + C$$

$$\int e^{ax+b} \, dx = \frac{1}{a}e^{ax+b} + C$$

2 $\int f(x)f'(x) \, dx = \frac{1}{2}[f(x)]^2 + C$

3 $\int [f(x)]^n f'(x) \, dx = \frac{1}{n+1}[f(x)]^{n+1} + C, n \neq -1$

4 $\int \frac{f'(x)}{f(x)} \, dx = \ln|f(x)| + C$

5 Integration by parts gives

$$\int v \frac{du}{dx} \, dx = uv - \int u \frac{dv}{dx} \, dx$$

6 $y = \int f(x) \, dx + C$ is called the general solution of the

differential equation $\frac{dy}{dx} = f(x)$, once the integration has been completed.

7 The differential equation $\frac{dy}{dx} = f(x) \, g(y)$ is called a first order equation with variables separable. Its general solution is

$$\int \frac{1}{g(y)} \, dy = \int f(x) \, dx + C$$

provided that $\frac{1}{g(y)}$ can be integrated with respect to y and that $f(x)$ can be integrated with respect to x.

Vectors

5

(a) The distance between P and Q is 200 m.

(b) The volume of the prism is $65 \, cm^3$.

(c) The time taken to run round the block was 184 s.

(d) A man walks 2 km on a bearing of 062°.

(e) A horizontal force of 3 N was applied at right angles to the length of the book.

(f) The velocity of the aeroplane is $800 \, km \, h^{-1}$ on a bearing 158°.

These six sentences give examples of everyday quantities. The first three (distance, volume, time) are called **scalar** quantities and they only need a single number or *scalar* ($200 \, m$, $65 \, cm^3$, $184 \, s$) to specify them precisely.

However in (d), (e) and (f) the quantities need both a magnitude ($2 \, km$, $3 \, N$, $800 \, km \, h^{-1}$) *and* a direction to specify them precisely. These are examples of **vector** quantities.

■ **A vector is a quantity that has both magnitude and a direction in space.**

5.1 Some definitions

Directed line segment

A vector can be represented by a **directed line segment** whose direction is that of the vector and whose length represents the magnitude of the vector.

For example, this directed line segment represents the vector $\overrightarrow{AB}$ where the arrow on the diagram represents the *direction* from A to B and the length of the line AB represents the *magnitude* of the vector.

Such a vector can be written as $\overrightarrow{AB}$ where A and B are the end-points of the directed line segment and the arrow above the letters indicates the direction of the vector. Alternatively, the vector is often denoted by a single lower case bold letter, for example, **p**.

When you need to use this alternative notation you should write the vector in *lower case* and *underline* it (i.e. p̲) since you will not be able to write it in bold print.

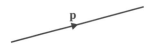

Displacement vector

A displacement is one of the most common types of vector. For example, a journey from P to Q of 90 km north east is a movement or displacement. You can write it as $\overrightarrow{PQ}$.

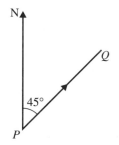

If the journey from P to Q is followed by a further journey from Q to R of 65 km south east, then the overall displacement is from P to R. The combination of the journey (displacement) from P to Q and the journey from Q to R results in the overall displacement of P to R:

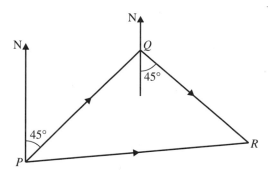

This is written as

$$\overrightarrow{PQ} + \overrightarrow{QR} = \overrightarrow{PR}$$

That is, the sum (or **resultant**) of the displacements $\overrightarrow{PQ}$ and $\overrightarrow{QR}$ is $\overrightarrow{PR}$.

Notice that if you want to add displacements, the end-point of one displacement vector must be the starting point of the next displacement vector. That is, $\overrightarrow{AC} + \overrightarrow{CQ} + \overrightarrow{QR} + \overrightarrow{RX} = \overrightarrow{AX}$, but $\overrightarrow{AB} + \overrightarrow{BE} + \overrightarrow{FH}$ can only be simplified to $\overrightarrow{AE} + \overrightarrow{FH}$ because the finishing point of the displacement $\overrightarrow{AE}$ is E and the starting point of the displacement $\overrightarrow{FH}$ is F, and these points are not the same.

Modulus of a vector

The **modulus** of a vector is its magnitude. The modulus of the vector $\overrightarrow{PQ}$ is written $|\overrightarrow{PQ}|$. The modulus of the vector **a** is written $|\mathbf{a}|$ or a.

If $\overrightarrow{PQ}$ is a velocity of $200 \, \text{km h}^{-1}$ north east, then

$$|\overrightarrow{PQ}| = 200 \, \text{km h}^{-1}$$

Equality of vectors

Two vectors are said to be equal if they have both the same magnitude and direction.

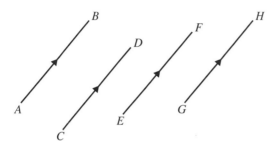

The four vectors $\overrightarrow{AB}$, $\overrightarrow{CD}$, $\overrightarrow{EF}$ and $\overrightarrow{GH}$ are each parallel to the others and all have the same magnitude. So the directed line segments used to represent them are parallel and of the same length. You can therefore write: $\overrightarrow{AB} = \overrightarrow{CD} = \overrightarrow{EF} = \overrightarrow{GH}$.

The zero vector

The **zero vector**, written **0**, has zero magnitude and indeterminate direction. Since the displacement vector $\overrightarrow{PQ}$ goes from P to Q and the displacement vector $\overrightarrow{QP}$ goes from Q to P you can write $\overrightarrow{PQ} + \overrightarrow{QP} = \mathbf{0}$, because a displacement from P to Q followed by a displacement from Q to P takes you back to where you started and so there is no overall displacement.

Negative vector

Since $\overrightarrow{PQ} + \overrightarrow{QP} = \mathbf{0}$, you can write $\overrightarrow{QP} = -\overrightarrow{PQ}$. That is, $\overrightarrow{QP}$ is the negative of the vector $\overrightarrow{PQ}$.

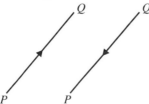

So the vector $\overrightarrow{QP}$ has the same magnitude as the vector $\overrightarrow{PQ}$ but its direction is exactly opposite to that of $\overrightarrow{PQ}$.

Example 1

Simplify: (a) $\overrightarrow{AB} + \overrightarrow{BC} - \overrightarrow{DC}$ (b) $\overrightarrow{AC} - \overrightarrow{BC} + \overrightarrow{BD}$.

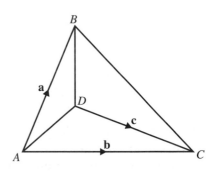

(a)
$$\overrightarrow{AB} + \overrightarrow{BC} - \overrightarrow{DC} = \overrightarrow{AC} - \overrightarrow{DC}$$
$$= \overrightarrow{AC} + \overrightarrow{CD}$$
$$= \overrightarrow{AD}$$

(b)
$$\overrightarrow{AC} - \overrightarrow{BC} + \overrightarrow{BD} = \overrightarrow{AC} + \overrightarrow{CB} + \overrightarrow{BD}$$
$$= \overrightarrow{AB} + \overrightarrow{BD}$$
$$= \overrightarrow{AD}$$

Example 2

If $\overrightarrow{AB} = \mathbf{a}$, $\overrightarrow{AC} = \mathbf{b}$ and $\overrightarrow{DC} = \mathbf{c}$, find, in terms of $\mathbf{a}$, $\mathbf{b}$ and $\mathbf{c}$:
(a) $\overrightarrow{BC}$ (b) $\overrightarrow{BD}$ (c) $\overrightarrow{AD}$.

(a)
$$\overrightarrow{BC} = \overrightarrow{BA} + \overrightarrow{AC}$$
$$= -\overrightarrow{AB} + \overrightarrow{AC}$$
$$= -\mathbf{a} + \mathbf{b}$$

(b)
$$\overrightarrow{BD} = \overrightarrow{BC} + \overrightarrow{CD}$$
$$= \overrightarrow{BC} - \overrightarrow{DC}$$
$$= (-\mathbf{a} + \mathbf{b}) - \mathbf{c}$$
$$= -\mathbf{a} + \mathbf{b} - \mathbf{c}$$

(c)
$$\overrightarrow{AD} = \overrightarrow{AC} + \overrightarrow{CD}$$
$$= \overrightarrow{AC} - \overrightarrow{DC}$$
$$= \mathbf{b} - \mathbf{c}$$

Exercise 5A

1 Simplify:

  (a) $\overrightarrow{AB} + \overrightarrow{BC} + \overrightarrow{CF}$ (b) $\overrightarrow{PQ} + \overrightarrow{ST} + \overrightarrow{QS} + \overrightarrow{TU}$

  (c) $\overrightarrow{LM} - \overrightarrow{PM} + \overrightarrow{QR}$ (d) $\overrightarrow{AC} - \overrightarrow{FC} - \overrightarrow{HF}$

  (e) $\overrightarrow{PQ} + \overrightarrow{QR} + \overrightarrow{QS} + \overrightarrow{RQ}$

2 In the figure $\overrightarrow{AB} = \mathbf{p}$, $\overrightarrow{DC} = \mathbf{r}$, $\overrightarrow{DB} = \mathbf{q}$.

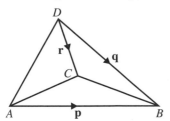

Find, in terms of $\mathbf{p}$, $\mathbf{q}$ and $\mathbf{r}$ expressions for

  (a) $\overrightarrow{BC}$ (b) $\overrightarrow{DA}$ (c) $\overrightarrow{AC}$.

3 In the figure $\overrightarrow{EA} = \mathbf{a}$, $\overrightarrow{AB} = \mathbf{b}$, $\overrightarrow{CD} = \mathbf{c}$, $\overrightarrow{DA} = \mathbf{d}$.

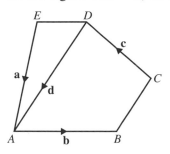

Find, in terms of **a**, **b**, **c** and **d**:
(a) $\overrightarrow{ED}$ (b) $\overrightarrow{BC}$ (c) $\overrightarrow{BE}$ (d) $\overrightarrow{DB}$.

5.2 More definitions and operations on vectors

Unit vector

A **unit vector** is a vector whose magnitude (modulus) is 1.

The vector **a** has magnitude $|\mathbf{a}| = a$. So a unit vector in the same direction as **a** is $\dfrac{\mathbf{a}}{a}$.

Scalar multiplication of a vector

Scalar multiplication means multiplying a vector by a scalar, that is, a number. The result is another vector.

Vectors such as $2\mathbf{a}$, $5\mathbf{a}$, $-3\mathbf{a}$ can be defined as $2\mathbf{a} = \mathbf{a} + \mathbf{a}$, $5\mathbf{a} = \mathbf{a} + \mathbf{a} + \mathbf{a} + \mathbf{a} + \mathbf{a}$, $-3\mathbf{a} = -\mathbf{a} - \mathbf{a} - \mathbf{a}$, and so on. Now $\mathbf{a} + \mathbf{a}$ is a displacement of **a** followed by another displacement of **a**. So $2\mathbf{a}$ is a vector in the same direction as **a** but which has twice the magnitude of **a**. Similarly, $5\mathbf{a}$ is in the same direction as **a** but has five times its magnitude. The vector $-3\mathbf{a}$ is in the opposite direction to **a** and has three times its magnitude. (Here the scalar is -3.)

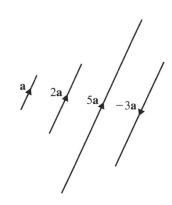

■ In general, the vector $\lambda\mathbf{a}$, where $\lambda > 0$, is in the same direction as **a** but its magnitude is λ times that of **a**.

Parallel vectors

From the above you should be able to see that if two vectors **a** and **b** are parallel then one is a scalar multiple of the other, that is: $\mathbf{a} = \lambda\mathbf{b}$.

If λ is positive then **a** is in the same direction as **b**. If λ is negative then **a** is in the opposite direction to **b**.

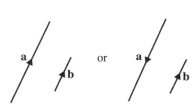

Adding vectors

Vectors are added in the same way as journeys.

It is clear that a journey from A to B followed by a journey from B to C is equivlent to a journey from A to C.

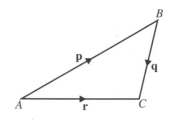

That is:
$$\overrightarrow{AB} + \overrightarrow{BC} = \overrightarrow{AC}$$

or
$$\mathbf{p} + \mathbf{q} = \mathbf{r}$$

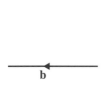 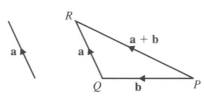

If $\mathbf{a}$ and $\mathbf{b}$ are two vectors $\overrightarrow{QR}$ and $\overrightarrow{PQ}$ then $\mathbf{a} + \mathbf{b}$ is the vector $\overrightarrow{PR}$. This is known as the **triangle law** for the addition of vectors. You must be sure that when you try to add two vectors $\mathbf{a}$ and $\mathbf{b}$ by the triangle law, the arrows of the line segments representing $\mathbf{a}$ and $\mathbf{b}$ on the diagram go in the same sense round the triangle. That is, they must both go round clockwise or both go round anticlockwise. You should also notice that the arrow of the line segment representing the resultant $\mathbf{a} + \mathbf{b}$ goes in the *opposite* sense to $\mathbf{a}$ and $\mathbf{b}$.

Another way to add the two vectors is to complete a parallelogram $PQRS$, where $\overrightarrow{RQ} = \mathbf{a}$ and $\overrightarrow{RS} = \mathbf{b}$. Remember that opposite sides of a parallelogram are equal and parallel. So $\overrightarrow{SP} = \overrightarrow{RQ} = \mathbf{a}$ and $\overrightarrow{QP} = \overrightarrow{RS} = \mathbf{b}$.

By using the triangle law in triangle PQR, as before

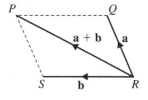

$$\overrightarrow{RQ} + \overrightarrow{QP} = \overrightarrow{RP}$$

that is:
$$\overrightarrow{RQ} + \overrightarrow{RS} = \overrightarrow{RP}$$

This is sometimes called the **parallelogram law** of addition.

When you use the parallelogram law of addition to add two vectors $\overrightarrow{RQ}$ and $\overrightarrow{RS}$ you should note that it is the *diagonal* $\overrightarrow{RP}$ of the parallelogram that represents the vector $\mathbf{a} + \mathbf{b}$.

The commutative law

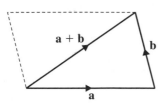 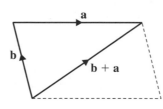

From the left-hand diagram you can see that the diagonal of the parallelogram represents $\mathbf{a} + \mathbf{b}$. From the right-hand diagram you can see that the diagonal represents $\mathbf{b} + \mathbf{a}$. But the vectors $\mathbf{a} + \mathbf{b}$ and $\mathbf{b} + \mathbf{a}$ are parallel and of equal magnitude. That is, they are equal.

- So: $$\mathbf{a} + \mathbf{b} = \mathbf{b} + \mathbf{a}$$

That is, *it does not matter which way round you add two vectors.*

This is known as the **commutative law**.

The associative law

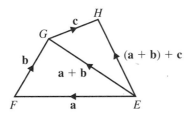 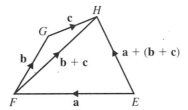

In the left-hand diagram, from $\triangle EFG$:
$$\overrightarrow{EG} = \mathbf{a} + \mathbf{b}$$

From $\triangle EGH$: $\qquad \overrightarrow{EH} = (\mathbf{a} + \mathbf{b}) + \mathbf{c}$

In the right-hand diagram, from $\triangle FGH$:
$$\overrightarrow{FH} = \mathbf{b} + \mathbf{c}$$

From $\triangle EFH$: $\qquad \overrightarrow{EH} = \mathbf{a} + (\mathbf{b} + \mathbf{c})$

- So: $\qquad (\mathbf{a} + \mathbf{b}) + \mathbf{c} = \mathbf{a} + (\mathbf{b} + \mathbf{c}) = \mathbf{a} + \mathbf{b} + \mathbf{c}$

That is, *it does not matter in which order you add vectors*

This is known as the **associative law**.

Subtracting vectors

If, once again, $\overrightarrow{RQ} = \mathbf{a}$ and $\overrightarrow{RS} = \mathbf{b}$ then:
$$\overrightarrow{SQ} = \overrightarrow{SR} + \overrightarrow{RQ}$$

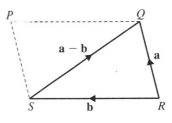

But $\qquad \overrightarrow{SR} = -\overrightarrow{RS} = -\mathbf{b}$

So $\qquad \overrightarrow{SQ} = -\mathbf{b} + \mathbf{a}$

or: $\qquad \overrightarrow{SQ} = \mathbf{a} - \mathbf{b}$

So the diagonal $\overrightarrow{SQ}$ of the parallelogram represents the vector $\mathbf{a} - \mathbf{b}$.

Non-parallel vectors

If $\mathbf{a}$ and $\mathbf{b}$ are *not* parallel and $\lambda\mathbf{a} + \mu\mathbf{b} = \alpha\mathbf{a} + \beta\mathbf{b}$, where λ, μ, α, β and scalars, then:
$$\lambda\mathbf{a} - \alpha\mathbf{a} = \beta\mathbf{b} - \mu\mathbf{b}$$

that is: $\qquad (\lambda - \alpha)\mathbf{a} = (\beta - \mu)\mathbf{b}$

Since **a** and **b** are not parallel then $(\lambda - \alpha)$ cannot be a scalar multiple of $(\beta - \mu)$. So the only way they can be equal is if $\lambda - \alpha = 0$ and $\beta - \mu = 0$.

That is: $$\lambda = \alpha \text{ and } \beta = \mu$$

■ $\lambda\mathbf{a} + \mu\mathbf{b} = \alpha\mathbf{a} + \beta\mathbf{b}$ where **a** and **b** are non-parallel $\Rightarrow \lambda = \alpha$ and $\mu = \beta$.

Example 3

$ABCD$ is a trapezium with AB parallel to DC.

$\overrightarrow{DA} = \mathbf{a}$, $\overrightarrow{AB} = 2\mathbf{b}$, $\overrightarrow{DM} = \overrightarrow{AB}$.

The point M is such that $DM : MC = 2 : 1$ and N is the mid-point of BC. Find, in terms of **a** or **b** or both **a** and **b**,

(a) $\overrightarrow{DC}$ (b) $\overrightarrow{AM}$ (c) $\overrightarrow{BC}$ (d) $\overrightarrow{MN}$

(a)
$$\overrightarrow{DM} = \overrightarrow{AB} = 2\mathbf{b}$$
$$DM : MC = 2 : 1$$

Thus:
$$\overrightarrow{MC} = \mathbf{b} \text{ and } \overrightarrow{DC} = 3\mathbf{b}$$

(b)
$$\overrightarrow{AM} = \overrightarrow{AD} + \overrightarrow{DM} = -\mathbf{a} + 2\mathbf{b}$$

(c)
$$\overrightarrow{BC} = \overrightarrow{BA} + \overrightarrow{AD} + \overrightarrow{DC}$$
$$= -\overrightarrow{AB} - \overrightarrow{DA} + \overrightarrow{DC}$$
$$= -2\mathbf{b} - \mathbf{a} + 3\mathbf{b}$$
$$= \mathbf{b} - \mathbf{a}$$

(d)
$$\overrightarrow{MN} = \overrightarrow{MC} + \overrightarrow{CN} = \mathbf{b} - \overrightarrow{NC}$$

Since N is the mid-point of BC,
$$\overrightarrow{NC} = \tfrac{1}{2}(\mathbf{b} - \mathbf{a})$$

So:
$$\overrightarrow{MN} = \mathbf{b} - \tfrac{1}{2}(\mathbf{b} - \mathbf{a})$$
$$= \mathbf{b} - \tfrac{1}{2}\mathbf{b} + \tfrac{1}{2}\mathbf{a}$$
$$= \tfrac{1}{2}\mathbf{b} + \tfrac{1}{2}\mathbf{a}$$

Example 4

The vectors **a** and **b** are not parallel and
$$(\lambda - \mu)\mathbf{a} + (\mu + 1)\mathbf{b} = 7\mathbf{a} - (\lambda + 2)\mathbf{b}$$
where λ and μ are scalars. Find the value of λ and the value of μ.

Since $$(\lambda - \mu)\mathbf{a} + (\mu + 1)\mathbf{b} = 7\mathbf{a} - (\lambda + 2)\mathbf{b}$$

and **a** and **b** are not parallel, then:
$$\lambda - \mu = 7 \qquad\qquad (1)$$
and $$\mu + 1 = -(\lambda + 2) \qquad\qquad (2)$$
(2) can be written $$\lambda + \mu = -3 \qquad\qquad (3)$$

So (1) + (3) gives: $2\lambda = 4 \Rightarrow \lambda = 2$

Substituting in (1) gives: $2 - \mu = 7 \Rightarrow \mu = -5$

Exercise 5B

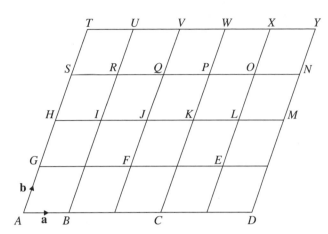

1 These two sets of lines are equally spaced and $\overrightarrow{AB} = \mathbf{a}$, $\overrightarrow{AG} = \mathbf{b}$. Find, in terms of $\mathbf{a}$ and $\mathbf{b}$:

(a) $\overrightarrow{AC}$ (b) $\overrightarrow{CD}$ (c) $\overrightarrow{FV}$ (d) $\overrightarrow{YM}$ (e) $\overrightarrow{UB}$

(f) $\overrightarrow{AF}$ (g) $\overrightarrow{IX}$ (h) $\overrightarrow{XN}$ (i) $\overrightarrow{WJ}$ (j) $\overrightarrow{AX}$

(k) $\overrightarrow{LV}$ (l) $\overrightarrow{EU}$ (m) $\overrightarrow{DT}$

2 State which of the following vectors are parallel to $2\mathbf{a} - 5\mathbf{b}$:

(a) $6\mathbf{a} - 15\mathbf{b}$ (b) $-2\mathbf{a} + 5\mathbf{b}$

(c) $\mathbf{a} - 3\mathbf{b}$ (d) $-\mathbf{a} + 2\mathbf{b}$

(e) $\frac{2}{5}\mathbf{a} - \mathbf{b}$

3 $ABCD$ is a parallelogram with $\overrightarrow{AB} = \mathbf{a}$ and $\overrightarrow{AD} = \mathbf{b}$. The point E is such that $\overrightarrow{DE} = 2\mathbf{b}$. Draw a sketch to illustrate this and express the vectors $\overrightarrow{AE}$, $\overrightarrow{AC}$ and $\overrightarrow{EC}$ in terms of $\mathbf{a}$ and $\mathbf{b}$.   [E]

4 In the triangle PQR, $\overrightarrow{PR} = \mathbf{a}$ and $\overrightarrow{PQ} = \mathbf{b}$. The point N lies on QR and is such that $QN : NR = 1 : 4$. Find, in terms of $\mathbf{a}$ and $\mathbf{b}$:

(a) $\overrightarrow{QR}$ (b) $\overrightarrow{RN}$ (c) $\overrightarrow{PN}$.

5 $ABCD$ is a trapezium with BC parallel to AD. If $\overrightarrow{AB} = \mathbf{a}$, $\overrightarrow{BC} = \mathbf{b}$, $BC = \frac{1}{3}AD$ and E is the mid-point of BC, find, in terms of $\mathbf{a}$ and $\mathbf{b}$:

(a) $\overrightarrow{CD}$ (b) $\overrightarrow{AE}$ (c) $\overrightarrow{DB}$ (d) $\overrightarrow{DE}$.

6 $ABCD$ is a parallelogram with $\overrightarrow{AB} = \mathbf{a}$ and $\overrightarrow{AD} = \mathbf{b}$. The point P lies on AD and is such that $AP : PD = 1 : 2$ and the point Q lies on BD and is such that $BQ : QD = 2 : 1$. Show that PQ is parallel to AC.

7 In the quadrilateral $OABC$, D is the mid-point of BC and G is the point on AD such that $AG : GD = 2 : 1$. Given that $\overrightarrow{OA} = \mathbf{a}$, $\overrightarrow{OB} = \mathbf{b}$ and $\overrightarrow{OC} = \mathbf{c}$, express $\overrightarrow{OD}$ and $\overrightarrow{OG}$ in terms of $\mathbf{a}$, $\mathbf{b}$ and $\mathbf{c}$. [E]

8 In the regular hexagon $PQRSTU$, $\overrightarrow{PQ} = \mathbf{a}$ and $\overrightarrow{QR} = \mathbf{b}$. Find in terms of $\mathbf{a}$ and $\mathbf{b}$:

(a) $\overrightarrow{PR}$ (b) $\overrightarrow{RS}$ (c) $\overrightarrow{RT}$ (d) $\overrightarrow{RP}$.

9 ABC is a triangle in which P and Q are the mid-points of AC and BC respectively. Prove that $\overrightarrow{BQ} + \overrightarrow{PQ} = \overrightarrow{AP}$.

10 In $\triangle OAB$, $\overrightarrow{OA} = 6\mathbf{a}$ and $\overrightarrow{OB} = 6\mathbf{b}$. The mid-point of OA is M and the point P lies on AB such that $AP : PB = 2 : 1$. The mid-point of OP is N.

(a) Calculate, in terms of $\mathbf{a}$ and $\mathbf{b}$, the vectors $\overrightarrow{AB}$, $\overrightarrow{OP}$ and $\overrightarrow{MN}$.

(b) Show that the area of the quadrilateral $AMNP$ is half the area of $\triangle OAB$.

The line AN produced meets OB at C.

(c) Given that $\overrightarrow{OC} = k\mathbf{b}$, find the value of k. [E]

11 Find the values of λ and μ given that $\mathbf{a}$ and $\mathbf{b}$ are not parallel:

(a) $5\mathbf{a} + \lambda\mathbf{b} = (6 - \mu)\mathbf{a} + 7\mathbf{b}$

(b) $(8 + \lambda)\mathbf{a} + (\mu - 2)\mathbf{b} = \mathbf{0}$

(c) $2\mathbf{a} + 3\mathbf{b} - (\lambda - 4)\mathbf{a} + (2 - \mu)\mathbf{b} = \mathbf{0}$

(d) $(2\lambda - 3)\mathbf{a} + 7\mathbf{b} = (5 - \lambda)\mathbf{a} + (2 - \mu)\mathbf{b}$

(e) $7\lambda\mathbf{a} + 5\lambda\mathbf{b} + 3\mu\mathbf{a} - \mu\mathbf{b} = 5\mathbf{a} + 2\mathbf{b}$

(f) $2\lambda\mathbf{a} + 3\lambda\mathbf{b} + 3\mu\mathbf{a} - 5\mu\mathbf{b} = 21\mathbf{b} - 5\mathbf{a}$

(g) $2\lambda\mathbf{a} + 3\mu\mathbf{b} = 7\mu\mathbf{a} + 11\lambda\mathbf{b} + 57\mathbf{a} + 6\mathbf{b}$

(h) $\lambda\mathbf{a} + 3\lambda\mathbf{b} + \mu\mathbf{b} = 2\mu\mathbf{a} + 8\mathbf{b} + 5\mathbf{a}$

12 In the figure, $\overrightarrow{OA} = \mathbf{a}$, $\overrightarrow{OB} = \mathbf{b}$ and C divides AB in the ratio $5 : 1$.

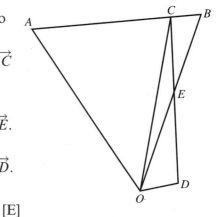

(a) Write down, in terms of $\mathbf{a}$ and $\mathbf{b}$, expressions for $\overrightarrow{AB}$, $\overrightarrow{AC}$ and $\overrightarrow{OC}$.

Given that $\overrightarrow{OE} = \lambda\mathbf{b}$, where λ is a scalar,

(b) write down, in terms of $\mathbf{a}$, $\mathbf{b}$ and λ, an expression for $\overrightarrow{CE}$.

Given that $\overrightarrow{OD} = \mu(\mathbf{b} - \mathbf{a})$, where μ is a scalar,

(c) write down, in terms of $\mathbf{a}$, $\mathbf{b}$, λ and μ, an expression for $\overrightarrow{ED}$.

Given also that E is the mid-point of CD,

(d) deduce the values of λ and μ. [E]

13 In the figure the points A and B are such that $\overrightarrow{OA} = \mathbf{a}$ and
$\overrightarrow{OB} = \mathbf{b}$. The point C lies on OB and is such that
$OC : CB = 2 : 1$. The point M is the mid-point of BC
and the point N is the mid-point of AC.
Find, in terms of $\mathbf{a}$, $\mathbf{b}$, or $\mathbf{a}$ and $\mathbf{b}$,
(a) $\overrightarrow{OC}$ (b) $\overrightarrow{OM}$ (c) $\overrightarrow{ON}$.

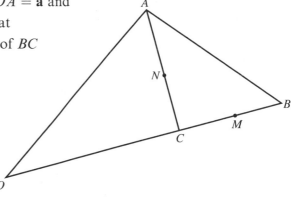

The point X lies on AM and is such that
$\overrightarrow{AX} = t\overrightarrow{AM}$, where $0 < t < 1$.
(d) Find $\overrightarrow{OX}$, in terms of $\mathbf{a}$, $\mathbf{b}$ and t.
The point Y lies on BN and is such that
$\overrightarrow{BY} = s\overrightarrow{BN}$, where $0 < s < 1$.
(e) Find $\overrightarrow{OY}$, in terms of $\mathbf{a}$, $\mathbf{b}$ and s. ·
P is the point of intersection of AM and BN.
(f) Find $\overrightarrow{OP}$, in terms of $\mathbf{a}$ and $\mathbf{b}$ only.
Q is the mid-point of AB.
(g) Show that C, P and Q are collinear and determine the
ratio $CP : CQ$. [E]

14 In the figure $\overrightarrow{OA} = \mathbf{a}$, $\overrightarrow{OB} = \mathbf{b}$, $\overrightarrow{OC} = 2\overrightarrow{OA}$, $\overrightarrow{OD} = 3\overrightarrow{OB}$, $\overrightarrow{BE} = \lambda\overrightarrow{BC}$,
$\overrightarrow{DE} = \mu\overrightarrow{DA}$ and $\overrightarrow{DF} = k\overrightarrow{DC}$, where λ, μ and k are constants.

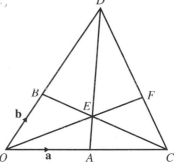

(a) Find, in terms of $\mathbf{a}$ and $\mathbf{b}$, (i) $\overrightarrow{BC}$ (ii) $\overrightarrow{DA}$ (iii) $\overrightarrow{DC}$.
(b) Find $\overrightarrow{OE}$, in terms of $\mathbf{a}$, $\mathbf{b}$ and λ.
(c) Show that $\overrightarrow{OE} = \mu\mathbf{a} + (3 - 3\mu)\mathbf{b}$.
(d) Using your answers to (b) and (c), show that
$\overrightarrow{OE} = \frac{1}{5}(4\mathbf{a} + 3\mathbf{b})$.
(e) Show that $\overrightarrow{OF} = 2k\mathbf{a} + (3 - 3k)\mathbf{b}$.
(f) Using your answers to (d) and (e), find $\overrightarrow{OF}$, in terms of $\mathbf{a}$
and $\mathbf{b}$ only. [E]

15 In the figure $\overrightarrow{OA} = \mathbf{a}$, $\overrightarrow{OB} = \mathbf{b}$, $3\overrightarrow{OC} = 2\overrightarrow{OA}$ and $4\overrightarrow{OD} = 7\overrightarrow{OB}$.

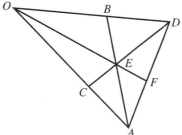

The line DC meets the line AB at E.
(a) Write down, in terms of $\mathbf{a}$ and $\mathbf{b}$, expressions for
(i) $\overrightarrow{AB}$ and (ii) $\overrightarrow{DC}$.

Given that $\overrightarrow{DE} = \lambda\overrightarrow{DC}$ and $\overrightarrow{EB} = \mu\overrightarrow{AB}$ where λ and μ are constants

(b) use $\triangle EBD$ to form an equation relating to **a**, **b**, λ and μ.

Hence

(c) show that $\lambda = \frac{9}{13}$

(d) find the exact value of μ

(e) express $\overrightarrow{OE}$ in terms of **a** and **b**.

The line OE produced meets the line AD at F.

Given that $\overrightarrow{OF} = k\overrightarrow{OE}$ where k is a constant and that $\overrightarrow{AF} = \frac{1}{10}(7\mathbf{b} - 4\mathbf{a})$

(f) find the value of k. [E]

16 In $\triangle OAB$, P is the mid-point of AB and Q is the point on OP such that $OQ = \frac{3}{4}OP$. Given that $\overrightarrow{OA} = \mathbf{a}$ and $\overrightarrow{OB} = \mathbf{b}$, find, in terms of **a** and **b**

(a) $\overrightarrow{AB}$ (b) $\overrightarrow{OP}$ (c) $\overrightarrow{OQ}$ (d) $\overrightarrow{AQ}$.

The point R on OB is such that $OR = kOB$, where $0 < k < 1$.

(e) Find, in terms of **a**, **b** and k, the vector $\overrightarrow{AR}$.

Given that AQR is a straight line

(f) find the ratio in which Q divides AR and the value of k. [E]

17 In the figure the points A and B are such that $\overrightarrow{OA} = \mathbf{a}$ and $\overrightarrow{OB} = \mathbf{b}$. The point M is the mid-point of OA. The point X is on OB such that X divides OB in the ratio $3:1$ and the point Y is on AX such that Y divides AX in the ratio $4:1$.

(a) Write down in terms of **a**, **b**, or **a** and **b**, expressions for $\overrightarrow{OM}$, $\overrightarrow{OX}$ and $\overrightarrow{OY}$.

(b) Show that $\overrightarrow{BY} = \frac{1}{5}(\mathbf{a} - 2\mathbf{b})$.

(c) Deduce that B, Y and M are collinear.

(d) Calculate the ratio $BY:YM$. [E]

18 In the figure $OE:EA = 1:2$, $AF:FB = 3:1$ and $OG:OB = 3:1$. The vector $\overrightarrow{OA} = \mathbf{a}$ and the vector $\overrightarrow{OB} = \mathbf{b}$.

Find, in terms of **a**, **b** or **a** and **b**, expressions for

(a) $\overrightarrow{OE}$ (b) $\overrightarrow{OF}$ (c) $\overrightarrow{EF}$ (d) $\overrightarrow{BG}$ (e) $\overrightarrow{FB}$ (f) $\overrightarrow{FG}$.

(g) Use your results in (c) and (f) to show that the points E, F and G are collinear and find the ratio $EF:FG$.

(h) Find $\overrightarrow{EB}$ and $\overrightarrow{AG}$ and hence prove that EB is parallel to AG. [E]

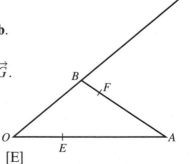

19 In the figure, $\overrightarrow{OA} = 5\mathbf{a}$, $\overrightarrow{AB} = 3\mathbf{b}$, $\overrightarrow{OC} = \frac{3}{2}\overrightarrow{OB}$ and $\overrightarrow{OD} = \frac{3}{5}\overrightarrow{OA}$.
The line DC meets AB at F.

(a) Write down, in terms of $\mathbf{a}$ and $\mathbf{b}$, expressions for
$\overrightarrow{OB}$, $\overrightarrow{OC}$ and $\overrightarrow{DC}$.

Given that $\overrightarrow{DF} = \lambda(\mathbf{a} + \mathbf{b})$ and $\overrightarrow{AF} = \mu\mathbf{b}$

(b) use the triangle ADF to form an equation
relating to $\mathbf{a}$, $\mathbf{b}$, λ and μ.

(c) Use your equation from part (b) to find the values of λ and μ.

(d) Deduce the ratios

(i) $AF : FB$ (ii) $DF : FC$.

A line is drawn through F parallel to AO to meet OB at G.

(e) Write down an expression, in terms of $\mathbf{a}$ and $\mathbf{b}$, for $\overrightarrow{OG}$.

[E]

20 In the figure, $\overrightarrow{OA} = \mathbf{a}$ and $\overrightarrow{OB} = \mathbf{b}$. The points P
and Q lie on OA and OB respectively, so that
$OP : PA = 2 : 1$ and $OQ : QB = 1 : 2$.

(a) Find, in terms of $\mathbf{a}$, $\mathbf{b}$ or $\mathbf{a}$ and $\mathbf{b}$,
the vectors

(i) $\overrightarrow{OP}$ (ii) $\overrightarrow{OQ}$ (iii) $\overrightarrow{PQ}$.

The point R is such that $OR : AR = 2 : 1$.

(b) Find, in terms of $\mathbf{a}$ and $\mathbf{b}$, the vector $\overrightarrow{RB}$.

(c) Show that RB is parallel to PQ and find the ratio
$RB : PQ$. The line segment QP is produced to a point S so
that $QP = PS$.

(d) Find, in terms of $\mathbf{a}$ and $\mathbf{b}$, the vectors $\overrightarrow{PS}$ and $\overrightarrow{AS}$.

(e) Show that the points B, A and S are collinear.   [E]

5.3 Position vectors

If you have a fixed origin O and a point A, then the vector $\overrightarrow{OA}$ is
defined to be the **position vector** of the point A. The line segment
representing $\overrightarrow{OA}$ starts at O and ends at A, so the vector $\overrightarrow{OA}$
uniquely defines the position of A.

Suppose you have two points A and B. The position vector of A
is $\overrightarrow{OA} = \mathbf{a}$. The position vector of B is $\overrightarrow{OB} = \mathbf{b}$. From the vector
triangle, you can see that the vector $\overrightarrow{AB}$ is $\mathbf{b} - \mathbf{a}$. Likewise, the
vector $\overrightarrow{BA}$ is $\mathbf{a} - \mathbf{b}$.

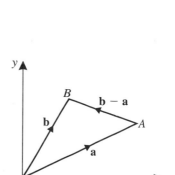

Position vector of the mid-point of a line

Let the position vector of a point A be $\mathbf{a}$ (that is, $\overrightarrow{OA} = \mathbf{a}$).

Let the position vector of B be $\mathbf{b}$.

Let M be the mid-point of AB.

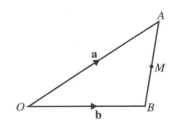

Then:
$$\overrightarrow{BA} = \mathbf{a} - \mathbf{b}$$

So:
$$\overrightarrow{OM} = \overrightarrow{OB} + \overrightarrow{BM}$$

$$= \overrightarrow{OB} + \tfrac{1}{2}\overrightarrow{BA}$$

$$= \mathbf{b} + \tfrac{1}{2}(\mathbf{a} - \mathbf{b})$$

$$= \tfrac{1}{2}(\mathbf{a} + \mathbf{b})$$

■ So the position vector of the mid-point of the line AB is $\tfrac{1}{2}(\mathbf{a} + \mathbf{b})$.

5.4 Cartesian components of a vector in two dimensions

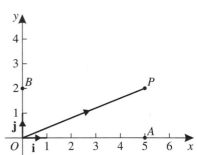

Consider the point $P(5, 2)$ referred to cartesian axes:

The vector $\overrightarrow{OP}$ is the position vector of P.

■ The vector $\mathbf{i}$ is defined to be a unit vector in the direction of x increasing parallel to the x-axis.
■ The vector $\mathbf{j}$ is defined to be a unit vector in the direction of y increasing parallel to the y-axis.

So if A is the point on the x-axis with coordinates $(5, 0)$ then A has position vector $5\mathbf{i}$. Similarly, $B(0, 2)$ has position vector $2\mathbf{j}$.

Now:
$$\overrightarrow{OP} = \overrightarrow{OA} + \overrightarrow{AP} = \overrightarrow{OA} + \overrightarrow{OB} = 5\mathbf{i} + 2\mathbf{j}$$

So the point P, with coordinates $(5, 2)$ has position vector $5\mathbf{i} + 2\mathbf{j}$.

This result can be generalised.

Any point $P(x, y)$ has position vector $\overrightarrow{OP}$. Now the distance $OA = x$ and $AP = y$.

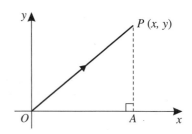

Also:
$$\overrightarrow{OP} = \overrightarrow{OA} + \overrightarrow{AP} = x\mathbf{i} + y\mathbf{j}$$

■ So the position vector of any point $P(x, y)$ is $x\mathbf{i} + y\mathbf{j}$.

x and y are called the **cartesian components** of $\overrightarrow{OP}$.

Now the length of the line OP which joins $(0, 0)$ to (x, y) is $\sqrt{(x^2 + y^2)}$.

■ So the modulus of $\overrightarrow{OP}$ is:

$$|\overrightarrow{OP}| = |x\mathbf{i} + y\mathbf{j}| = \sqrt{(x^2 + y^2)}$$

Since a unit vector in the direction of $\overrightarrow{OP}$ is defined as $\dfrac{\overrightarrow{OP}}{|\overrightarrow{OP}|}$, a

unit vector in the direction of $x\mathbf{i} + y\mathbf{j}$ is $\dfrac{x\mathbf{i} + y\mathbf{j}}{\sqrt{(x^2 + y^2)}}$.

Example 5
If $\mathbf{a} = 2\mathbf{i} + 7\mathbf{j}$, find $|\mathbf{a}|$.

$$|\mathbf{a}| = \sqrt{(2^2 + 7^2)} = \sqrt{53}.$$

Example 6
Given that $\mathbf{a} = -3\mathbf{i} + 2\mathbf{j}$ and $\mathbf{b} = 4\mathbf{i} - 7\mathbf{j}$, find: (a) $\mathbf{a} + \mathbf{b}$ (b) $\mathbf{a} - \mathbf{b}$
(c) $|\mathbf{a} - \mathbf{b}|$.

(a) $$\mathbf{a} + \mathbf{b} = (-3\mathbf{i} + 2\mathbf{j}) + (4\mathbf{i} - 7\mathbf{j}) = \mathbf{i} - 5\mathbf{j}$$

(b) $$\mathbf{a} - \mathbf{b} = (-3\mathbf{i} + 2\mathbf{j}) - (4\mathbf{i} - 7\mathbf{j})$$
$$= -3\mathbf{i} + 2\mathbf{j} - 4\mathbf{i} + 7\mathbf{j}$$
$$= -7\mathbf{i} + 9\mathbf{j}$$

(c) $$|\mathbf{a} - \mathbf{b}| = |-7\mathbf{i} + 9\mathbf{j}| = \sqrt{(49 + 81)} = \sqrt{130}$$

Example 7
Find a unit vector in the direction of $-2\mathbf{i} + 5\mathbf{j}$.

$$|-2\mathbf{i} + 5\mathbf{j}| = \sqrt{(2^2 + 5^2)} = \sqrt{29}$$

So a unit vector in the direction of $-2\mathbf{i} + 5\mathbf{j}$ is

$$\frac{1}{\sqrt{29}}(-2\mathbf{i} + 5\mathbf{j}) = -\frac{2}{\sqrt{29}}\mathbf{i} + \frac{5}{\sqrt{29}}\mathbf{j}$$

5.5 Cartesian coordinates in three dimensions

In Book P1 you were introduced to the ideas of coordinate geometry. Coordinate geometry is the study of the geometry of points, lines, curves and planes using algebraic methods. In Book P1 you looked at lines or curves in two dimensions (2D) only. Now we have a look at some figures in three dimensions (3D).

If two axes at right angles to each other are laid on a plane, then you can identify any point in the plane, using coordinates that show how far the point is from each axis. One of these axes is usually called the *x*-axis, the other is called the *y*-axis and the point where the axes meet is called the origin and is labelled *O*.

A similar set-up can be used in three dimensions by adding a third axis, the *z*-axis, at right angles to both the *x*- and *y*-axes.

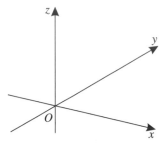

To define a point *P* in space, you must state the perpendicular distance of the point from the *yz*-plane, its perpendicular distance from the *xz*-plane and its perpendicular distance from the *xy*-plane. This defines the position of the point uniquely.

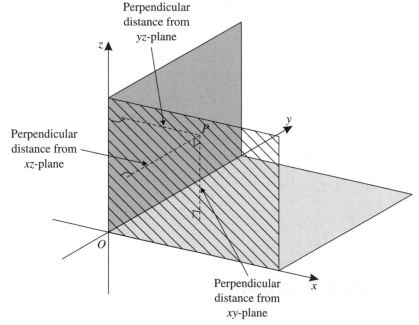

As in two dimensions, part of each axis is positive and the rest is negative. The part of the *x*-axis from *O* in the direction of the arrow in the diagram is the positive *x*-axis and the rest is the negative *x*-axis; the part of the *y*-axis from *O* in the direction of the arrow is the positive *y*-axis and the rest is the negative *y*-axis; and the part of the *z*-axis from *O* in the direction of the arrow is the positive *z*-axis and the rest is the negative *z*-axis.

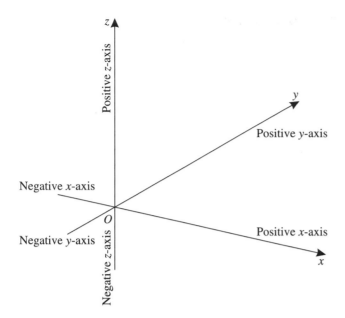

The perpendicular distance of a point in space from the yz-plane is called the **x-coordinate**, the perpendicular distance of the point from the xz-plane is called the **y-coordinate** and the perpendicular distance of the point from the xy-plane is called the **z-coordinate**. The coordinates of a point are written (x, y, z) where the first number is the x-coordinate, the second number is the y-coordinate and the third number is the z-coordinate. This is the point $P(2, 5, 4)$:

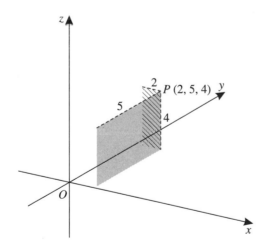

It is usual to define the directions of the coordinate axes like this:

(i) Oy is a 90° rotation *anticlockwise* from Ox in the xy-plane when viewed from the z-positive side of that plane.

(ii) Oz is a 90° rotation *anticlockwise* from Oy in the yz-plane when viewed from the x-positive side of that plane.

(iii) Ox is a 90° rotation *anticlockwise* from Oz in the xz-plane when viewed from the y-positive side of that plane.

Finding the distance between two points

Suppose you want to find the distance AB between the points $A(2, 1, 3)$ and $B(4, 5, 8)$.

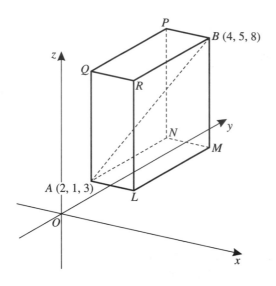

Draw a cuboid that has AB as a diagonal and has each of its faces parallel to the coordinate planes. In the diagram,

(i) the faces $LMBR$ and $ANPQ$ are parallel to the yz-plane,

(ii) the faces $MBPN$ and $LRQA$ are parallel to the xz-plane,

(iii) the faces $LMNA$ and $RBPQ$ are parallel to the xy-plane.

So:
$$AL = NM = 4 - 2 = 2$$
$$LM = AN = 5 - 1 = 4$$
$$MB = AQ = 8 - 3 = 5$$

In $\triangle AMB$:
$$AB^2 = AM^2 + MB^2$$
$$AB^2 = AM^2 + 5^2$$

In $\triangle ALM$:
$$AM^2 = AL^2 + LM^2$$
$$AM^2 = 2^2 + 4^2$$

So:
$$AB^2 = (2^2 + 4^2) + 5^2$$
$$= 4 + 16 + 25$$

and
$$AB = \sqrt{45}$$

To find a formula that can always be used to find the distance between two points, you can generalise the process by using two points with coordinates (x_1, y_1, z_1) and (x_2, y_2, z_2).

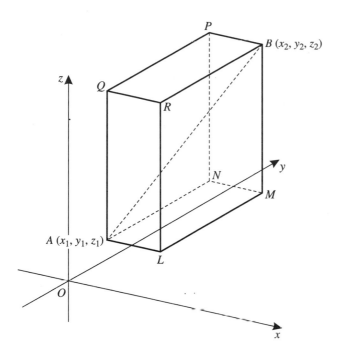

As before, draw the cuboid with AB as diagonal and with each face parallel to the coordinate planes.

So:
$$AL = NM = x_2 - x_1$$
$$LM = AN = y_2 - y_1$$
$$MB = AQ = z_2 - z_1$$

In $\triangle AMB$:
$$AB^2 = AM^2 + MB^2$$
$$AB^2 = AM^2 + (z_2 - z_1)^2$$

In $\triangle ALM$:
$$AM^2 = AL^2 + LM^2$$
$$AM^2 = (x_2 - x_1)^2 + (y_2 - y_1)^2$$

So:
$$AB^2 = [(x_2 - x_1)^2 + (y_2 - y_1)^2] + (z_2 - z_1)^2$$

or

■
$$AB = \sqrt{[(x_2 - x_1)^2 + (y_2 - y_1)^2 + (z_2 - z_1)^2]}$$

Example 8
Find the distance between the points $A(1, 3, 5)$ and $B(2, 6, 7)$.

Using the formula:

$$AB = \sqrt{[(x_2 - x_1)^2 + (y_2 - y_1)^2 + (z_2 - z_1)^2]}$$
$$= \sqrt{[(2 - 1)^2 + (6 - 3)^2 + (7 - 5)^2]}$$
$$= \sqrt{(1^2 + 3^2 + 2^2)}$$
$$= \sqrt{(1 + 9 + 4)}$$
$$= \sqrt{14}$$
$$= 3.74 \text{ (3 s.f.)}$$

Example 9
Find the distance between the points $A(6, -1, -3)$ and $B(2, 4, -1)$.

Using the formula:

$$AB = \sqrt{[(x_2 - x_1)^2 + (y_2 - y_1)^2 + (z_2 - z_1)^2]}$$
$$= \sqrt{[(2 - 6)^2 + (4 + 1)^2 + (-1 + 3)^2]}$$
$$= \sqrt{[(-4)^2 + 5^2 + 2^2]}$$
$$= \sqrt{(16 + 25 + 4)}$$
$$= \sqrt{45}$$
$$= 6.71 \text{ (3 s.f.)}$$

Example 10
The length of the line joining $A(3, 1, 4)$ to $B(1, t, -2)$ is 7. Calculate the two possible values of t.

By the formula:

$$AB^2 = (3 - 1)^2 + (1 - t)^2 + (4 + 2)^2$$
$$= 2^2 + (1 - 2t + t^2) + 6^2$$
$$= 4 + 1 - 2t + t^2 + 36$$
$$= t^2 - 2t + 41$$

But if $AB = 7$ then $AB^2 = 49$

So:
$$t^2 - 2t + 41 = 49$$
$$t^2 - 2t - 8 = 0$$
$$(t + 2)(t - 4) = 0$$
$$t = -2 \text{ or } 4$$

5.6 Cartesian components of a vector in three dimensions

If you assume that the vectors **i** and **j** are as previously defined and that the vector **k** is defined to be a unit vector in the direction of z increasing parallel to the z-axis, similar results apply in three dimensions as in two.

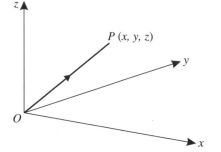

■ If P has coordinates (x, y, z)

then: $$\overrightarrow{OP} = x\mathbf{i} + y\mathbf{j} + z\mathbf{k}$$

■ The modulus of $\overrightarrow{OP}$ is $\sqrt{(x^2 + y^2 + z^2)}$.

■ A unit vector in the direction of $\overrightarrow{OP}$ is $\dfrac{x\mathbf{i} + y\mathbf{j} + z\mathbf{k}}{\sqrt{(x^2 + y^2 + z^2)}}$.

■ If A has coordinates (a_1, a_2, a_3) and B has coordinates (b_1, b_2, b_3) then:

$$\overrightarrow{OA} = a_1\mathbf{i} + a_2\mathbf{j} + a_3\mathbf{k}$$

$$\overrightarrow{OB} = b_1\mathbf{i} + b_2\mathbf{j} + b_3\mathbf{k}$$

$$\overrightarrow{AB} = (b_1 - a_1)\mathbf{i} + (b_2 - a_2)\mathbf{j} + (b_3 - a_3)\mathbf{k}$$

$$|\overrightarrow{AB}| = \sqrt{[(b_1 - a_1)^2 + (b_2 - a_2)^2 + (b_3 - a_3)^2]}$$

$$= \text{distance between the points } A \text{ and } B.$$

Example 11
Relative to an origin O, the points A and B have position vectors $3\mathbf{i} - 4\mathbf{j} + 6\mathbf{k}$ and $2\mathbf{i} - \mathbf{j} - \mathbf{k}$ respectively. Find the distance between A and B.

$$\overrightarrow{AB} = \overrightarrow{OB} - \overrightarrow{OA} = (2 - 3)\mathbf{i} + (-1 + 4)\mathbf{j} + (-1 - 6)\mathbf{k}$$

$$= -\mathbf{i} + 3\mathbf{j} - 7\mathbf{k}$$

Distance between A and $B = |\overrightarrow{AB}| = \sqrt{[(-1)^2 + 3^2 + (-7)^2]}$

$$= \sqrt{59}$$

Exercise 5C

1 Find the modulus of:

(a) $-5\mathbf{i} + 12\mathbf{j}$

(b) $-24\mathbf{i} + 7\mathbf{j}$

(c) $-\mathbf{i} - 5\mathbf{j}$

(d) $3\mathbf{i} - 2\mathbf{j} + 6\mathbf{k}$

(e) $-2\mathbf{i} - 2\mathbf{j} + \mathbf{k}$

(f) $5\mathbf{i} - 2\mathbf{j} - 5\mathbf{k}$

2 Find a unit vector in the direction of:

    (a) $-5\mathbf{i} + 12\mathbf{j}$ (b) $-\mathbf{i} - 5\mathbf{j}$

    (c) $4\mathbf{i} - 7\mathbf{j}$ (d) $3\mathbf{i} - 2\mathbf{j} + 6\mathbf{k}$

    (e) $8\mathbf{i} - \mathbf{j} - 4\mathbf{k}$ (f) $7\mathbf{i} - 5\mathbf{j} + \mathbf{k}$

3 Find a vector of magnitude 27 units which is parallel to $3\mathbf{i} + 4\mathbf{j}$.

4 Find the distance between $A(1, 2, 5)$ and $B(2, 3, 4)$.

5 Find the distance between $A(3, 1, 5)$ and $B(1, 7, 3)$.

6 Find the distance between $A(3, -2, -4)$ and $B(6, 1, 0)$.

7 Find the distance between $A(-1, -3, 5)$ and $B(2, -5, 4)$.

8 Find the distance between $A(-3, 2, -6)$ and $B(-4, -3, 7)$.

9 The point A has coordinates $(2, 1, 7)$ and the point B has coordinates $(3, t, 4)$. The distance AB is 7. Calculate the two possible values of t.

10 Find a vector of magnitude 3 units which is parallel to $\mathbf{i} - 3\mathbf{j} + 2\mathbf{k}$.

11 Given that $\mathbf{a} = \mathbf{i} + 2\mathbf{j} - 3\mathbf{k}$ and $\mathbf{b} = 2\mathbf{i} - 4\mathbf{j} - 5\mathbf{k}$, find

    (a) $\mathbf{a} + \mathbf{b}$ (b) $\mathbf{a} - \mathbf{b}$ (c) $|\mathbf{a} - \mathbf{b}|$ (d) $|-\mathbf{a} + 2\mathbf{b}|$.

12 Given that $\mathbf{a} = 2\mathbf{i} - 6\mathbf{j} + 3\mathbf{k}$ and $\mathbf{b} = -\mathbf{i} + 7\mathbf{j} - 5\mathbf{k}$, find

    (a) $|\mathbf{a} - \mathbf{b}|$ (b) $|2\mathbf{a} + 3\mathbf{b}|$

    (c) a unit vector in the direction of $2\mathbf{a} - \mathbf{b}$.

13 The point P has position vector $2\mathbf{i} - 4\mathbf{j} + 5\mathbf{k}$ and $\overrightarrow{PQ} = 3\mathbf{i} + 6\mathbf{j} - 2\mathbf{k}$. Find the position vector of the point Q.

14 Given that $\mathbf{a} = 2\lambda\mathbf{i} + 3\mathbf{j} - \lambda\mathbf{k}$ and that $|\mathbf{a}| = 5$ find the possible values of λ.

15 Given that $\mathbf{a} = 2\lambda\mathbf{i} - \mathbf{j} + 3\lambda\mathbf{k}$ and that $|\mathbf{a}| = \sqrt{27}$, find the possible values of λ.

16 Given that $\mathbf{a} = 3\mathbf{i} - 2\lambda\mathbf{j} + 5\lambda\mathbf{k}$ and that $|\mathbf{a}| = \sqrt{67}$, find the possible values of λ.

17 The point A has coordinates $(5, 3, -3)$ and the point B has coordinates $(-t, t, 3t)$.

    (a) Find an expression for AB^2.

    (b) Find the value of t that makes AB^2 a minimum.

    (c) Calculate the minimum value of AB^2.

18 The point A has coordinates $(3, 2t, 1)$ and the point B has coordinates $(-2t, 5, t)$.

    (a) Find an expression for AB^2.

    (b) Find the value of t that makes AB^2 a minimum.

    (c) Calculate the minimum value of AB^2.

19 The point A has coordinates $(3 + t, 2, 6 - t)$ and the point B has coordinates $(5, 1 + t, -3)$.

    (a) Find an expression for AB^2.

    (b) Find the value of t that makes AB^2 a minimum.

    (c) Calculate the minimum value of AB^2.

20 The point A has coordinates $(3, 0, 0)$, the point B has coordinates $(0, 3, 0)$ and the point C has coordinates $(0, 0, 7)$. Find, to $0.1°$, the size of the angle between the planes OAB and ABC, where O is the origin.

5.7 The scalar product of two vectors

The **scalar** (or **dot**) **product** of two vectors **a** and **b** is defined as $|\mathbf{a}||\mathbf{b}| \cos\theta$ where θ is the angle between **a** and **b**. The result of this calculation is a *scalar*, which is why it is called the scalar product.

You must note very carefully that this formula only works when the directions of the vectors **a** and **b** are either both towards their point of intersection or both away from their point of intersection and θ is the angle between their directions.

The scalar product is written **a.b** and you read it as 'a dot b'.

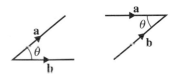

■ $$\mathbf{a}.\mathbf{b} = |\mathbf{a}||\mathbf{b}| \cos\theta$$

Notice the following properties of the scalar product:

1. For two perpendicular vectors:

$$\mathbf{a}.\mathbf{b} = 0$$

In particular, $$\mathbf{i}.\mathbf{j} = \mathbf{j}.\mathbf{k} = \mathbf{k}.\mathbf{i} = 0$$

If **a** and **b** are perpendicular

then: $$\mathbf{a}.\mathbf{b} = |\mathbf{a}||\mathbf{b}| \cos 90°$$

But $\cos 90° = 0$

So: $$\mathbf{a}.\mathbf{b} = 0$$

2. $$\mathbf{a}.\mathbf{a} = a^2$$

In particular, $\mathbf{i}.\mathbf{i} = \mathbf{j}.\mathbf{j} = \mathbf{k}.\mathbf{k} = 1$

$$\mathbf{a}.\mathbf{a} = |\mathbf{a}||\mathbf{a}| \cos\theta$$

But since they are the same vector the angle between them is zero.

So $$\mathbf{a}.\mathbf{a} = |\mathbf{a}||\mathbf{a}| \cos 0 = a.a.1 = a^2$$

3.
$$\mathbf{a} \cdot \mathbf{b} = \mathbf{b} \cdot \mathbf{a}$$

$$\mathbf{a} \cdot \mathbf{b} = |\mathbf{a}||\mathbf{b}| \cos \theta = |\mathbf{b}||\mathbf{a}| \cos \theta = \mathbf{b} \cdot \mathbf{a}$$

4.
$$\lambda(\mathbf{a} \cdot \mathbf{b}) = (\lambda \mathbf{a}) \cdot \mathbf{b} = \mathbf{a} \cdot (\lambda \mathbf{b})$$

$$\lambda(\mathbf{a} \cdot \mathbf{b}) = \lambda(|\mathbf{a}||\mathbf{b}| \cos \theta) = \lambda|\mathbf{a}||\mathbf{b}| \cos \theta$$

Since
$$|\lambda \mathbf{a}| = \lambda|\mathbf{a}| \text{ and } |\lambda \mathbf{b}| = \lambda|\mathbf{b}|$$

then:
$$(\lambda \mathbf{a}) \cdot \mathbf{b} = |\lambda \mathbf{a}||\mathbf{b}| \cos \theta = \lambda|\mathbf{a}||\mathbf{b}| \cos \theta = \lambda(\mathbf{a} \cdot \mathbf{b})$$

You can show that $\lambda(\mathbf{a} \cdot \mathbf{b}) = \mathbf{a} \cdot (\lambda \mathbf{b})$ in a similar way.

5.
$$\mathbf{a} \cdot (\mathbf{b} + \mathbf{c}) = \mathbf{a} \cdot \mathbf{b} + \mathbf{a} \cdot \mathbf{c}$$

In the diagram
$$\mathbf{d} = \mathbf{b} + \mathbf{c}$$

So
$$\mathbf{a} \cdot (\mathbf{b} + \mathbf{c}) = \mathbf{a} \cdot \mathbf{d}$$

Now
$$PQ + QR = PR$$

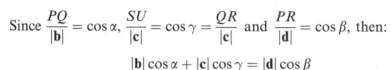

Since $\dfrac{PQ}{|\mathbf{b}|} = \cos \alpha$, $\dfrac{SU}{|\mathbf{c}|} = \cos \gamma = \dfrac{QR}{|\mathbf{c}|}$ and $\dfrac{PR}{|\mathbf{d}|} = \cos \beta$, then:

$$|\mathbf{b}| \cos \alpha + |\mathbf{c}| \cos \gamma = |\mathbf{d}| \cos \beta$$

Multiply through by $|\mathbf{a}|$:

$$|\mathbf{a}||\mathbf{b}| \cos \alpha + |\mathbf{a}||\mathbf{c}| \cos \gamma = |\mathbf{a}||\mathbf{d}| \cos \beta$$

that is:
$$\mathbf{a} \cdot \mathbf{b} + \mathbf{a} \cdot \mathbf{c} = \mathbf{a} \cdot \mathbf{d}$$

or:
$$\mathbf{a} \cdot \mathbf{b} + \mathbf{a} \cdot \mathbf{c} = \mathbf{a} \cdot (\mathbf{b} + \mathbf{c})$$

The scalar product in terms of cartesian coordinates

If the vectors $\mathbf{a}$ and $\mathbf{b}$ are given in terms of their cartesian components as $\mathbf{a} = x_1\mathbf{i} + y_1\mathbf{j} + z_1\mathbf{k}$ and $\mathbf{b} = x_2\mathbf{i} + y_2\mathbf{j} + z_2\mathbf{k}$, then

$$\begin{aligned}
\mathbf{a} \cdot \mathbf{b} &= (x_1\mathbf{i} + y_1\mathbf{j} + z_1\mathbf{k}) \cdot (x_2\mathbf{i} + y_2\mathbf{j} + z_2\mathbf{k}) \\
&= x_1 x_2 \mathbf{i} \cdot \mathbf{i} + x_1 y_2 \mathbf{i} \cdot \mathbf{j} + x_1 z_2 \mathbf{i} \cdot \mathbf{k} \\
&\quad + y_1 x_2 \mathbf{j} \cdot \mathbf{i} + y_1 y_2 \mathbf{j} \cdot \mathbf{j} + y_1 z_2 \mathbf{j} \cdot \mathbf{k} \\
&\quad + z_1 x_2 \mathbf{k} \cdot \mathbf{i} + z_1 y_2 \mathbf{k} \cdot \mathbf{j} + z_1 z_2 \mathbf{k} \cdot \mathbf{k}
\end{aligned}$$

Now since $\mathbf{i} \cdot \mathbf{i} = |\mathbf{i}||\mathbf{i}| \cos 0 = 1$, and likewise $\mathbf{j} \cdot \mathbf{j}$ and $\mathbf{k} \cdot \mathbf{k}$, and since $\mathbf{i} \cdot \mathbf{j} = |\mathbf{i}||\mathbf{j}| \cos 90° = 0$, and likewise $\mathbf{i} \cdot \mathbf{k} = \mathbf{j} \cdot \mathbf{i} = \mathbf{j} \cdot \mathbf{k} = \mathbf{k} \cdot \mathbf{i} = \mathbf{k} \cdot \mathbf{j} = 0$, then

∎
$$\mathbf{a} \cdot \mathbf{b} = x_1 x_2 + y_1 y_2 + z_1 z_2$$

Putting the two results for the scalar product together you get:

∎
$$x_1 x_2 + y_1 y_2 + z_1 z_2 = |\mathbf{a}||\mathbf{b}| \cos \theta$$

Example 12

Find the angle between the vectors $\mathbf{a} = 2\mathbf{i} - 3\mathbf{j} + \mathbf{k}$ and $\mathbf{b} = -\mathbf{i} + 5\mathbf{j} + 4\mathbf{k}$, giving your answer to the nearest tenth of a degree.

$\mathbf{a} \cdot \mathbf{b} = |\mathbf{a}||\mathbf{b}| \cos \theta$, where θ is the required angle

$$|\mathbf{a}| = \sqrt{(4 + 9 + 1)} = \sqrt{14}$$
$$|\mathbf{b}| = \sqrt{(1 + 25 + 16)} = \sqrt{42}$$

Also
$$\mathbf{a} \cdot \mathbf{b} = (2\mathbf{i} - 3\mathbf{j} + \mathbf{k}) \cdot (-\mathbf{i} + 5\mathbf{j} + 4\mathbf{k})$$
$$= -2 - 15 + 4$$
$$= -13$$

So:
$$\sqrt{14}\sqrt{42} \cos \theta = -13$$

$$\cos \theta = \frac{-13}{\sqrt{588}} = -0.5361$$

and:
$$\theta = 122.4°$$

Example 13

The points A, B and C have position vectors $\mathbf{a} = 4\mathbf{i} - \mathbf{k}$, $\mathbf{b} = \mathbf{i} + 2\mathbf{j} - 2\mathbf{k}$ and $\mathbf{c} = 3\mathbf{i} + 3\mathbf{j} - 6\mathbf{k}$ respectively referred to an origin O. Show that $\angle ABC = 90°$.

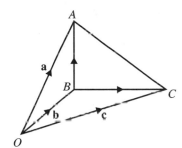

To use the scalar product you must have both vectors $\overrightarrow{BC}$ and $\overrightarrow{BA}$ in the directions away from the angle: Then:

$$\overrightarrow{BC} = \overrightarrow{OC} - \overrightarrow{OB} = \mathbf{c} - \mathbf{b} = 2\mathbf{i} + \mathbf{j} - 4\mathbf{k}$$
$$\overrightarrow{BA} = \overrightarrow{OA} - \overrightarrow{OB} = \mathbf{a} - \mathbf{b} = 3\mathbf{i} - 2\mathbf{j} + \mathbf{k}$$

So:
$$\overrightarrow{BC} \cdot \overrightarrow{BA} = (2 \times 3) + (1 \times -2) + (-4 \times 1)$$
$$= 6 - 2 - 4 = 0$$

Thus $\angle ABC = 90°$.

The projection of a vector on to another vector

You have already seen that a vector can be expressed in terms of its cartesian components. This is a special case. In fact, you can find the component of a vector in *any* direction. Remember that a component of a vector is a scalar quantity not a vector quantity.

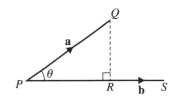

Let θ be the angle between $\mathbf{a}$ and $\mathbf{b}$ where $\overrightarrow{PQ} = \mathbf{a}$ and $\overrightarrow{PS} = \mathbf{b}$. Consider the vector $\mathbf{a}$ and its **projection** PR in the direction of $\mathbf{b}$.

Now
$$PR = PQ \cos \theta$$

But
$$\mathbf{a} \cdot \mathbf{b} = (PQ)(PS) \cos \theta$$

So:
$$PQ \cos \theta = \frac{\mathbf{a} \cdot \mathbf{b}}{PS}$$

But *PS* is the magnitude of **b**:

$$PS = |\mathbf{b}|$$

■ So the projection of **a** on **b** is $\dfrac{\mathbf{a} \cdot \mathbf{b}}{|\mathbf{b}|}$

Example 14

Find the projection of $\mathbf{a} = 2\mathbf{i} + \mathbf{j} - 5\mathbf{k}$ in the direction of $\mathbf{b} = -\mathbf{i} + 7\mathbf{j} + 2\mathbf{k}$.

$\mathbf{a} \cdot \mathbf{b} = -2 + 7 - 10 = -5$

$|\mathbf{b}| = \sqrt{(1 + 49 + 4)} = \sqrt{54}$

So the projection is $\dfrac{-5}{\sqrt{54}} = \dfrac{-5}{3\sqrt{6}}$

Notice that in this example the projection is negative. This means that the angle θ between the vectors **a** and **b** is obtuse, so the projection is in the *opposite* direction to **b**.

Exercise 5D

1 Find $\mathbf{a} \cdot \mathbf{b}$ where:
   (a) $\mathbf{a} = 2\mathbf{i} - \mathbf{j} + 5\mathbf{k}, \mathbf{b} = -3\mathbf{i} + 5\mathbf{j} - 7\mathbf{k}$
   (b) $\mathbf{a} = 5\mathbf{i} - \mathbf{k}, \mathbf{b} = -2\mathbf{i} + 6\mathbf{j} - 2\mathbf{k}$
   (c) $\mathbf{a} = -3\mathbf{i} + 4\mathbf{j} + 6\mathbf{k}, \mathbf{b} = \mathbf{k} - 2\mathbf{i}$
   (d) $\mathbf{a} = 6\mathbf{i} - 3\mathbf{k}, \mathbf{b} = 4\mathbf{j} + 7\mathbf{k}$
   (e) $\mathbf{a} = -5\mathbf{i} + 7\mathbf{j} - 8\mathbf{k}, \mathbf{b} = 2\mathbf{i} - 5\mathbf{j} + 11\mathbf{k}$

2 Find, in degrees to 1 decimal place, the angle between the vectors **a** and **b** where:
   (a) $\mathbf{a} = 2\mathbf{i} - \mathbf{j} - 2\mathbf{k}, \mathbf{b} = 3\mathbf{i} + \mathbf{j} - 5\mathbf{k}$
   (b) $\mathbf{a} = -\mathbf{i} + 4\mathbf{j} - 2\mathbf{k}, \mathbf{b} = 5\mathbf{i} + \mathbf{j} - 3\mathbf{k}$
   (c) $\mathbf{a} = -2\mathbf{i} + 6\mathbf{j} - \mathbf{k}, \mathbf{b} = \mathbf{i} - \mathbf{j} - 2\mathbf{k}$
   (d) $\mathbf{a} = -\mathbf{i} - 2\mathbf{j} + \mathbf{k}, \mathbf{b} = 3\mathbf{i} - 4\mathbf{k}$
   (e) $\mathbf{a} = \mathbf{i} + 3\mathbf{j} + \mathbf{k}, \mathbf{b} = 3\mathbf{i} - \mathbf{j} - \mathbf{k}$
   (f) $\mathbf{a} = -5\mathbf{i} - 2\mathbf{j} - \mathbf{k}, \mathbf{b} = 2\mathbf{i} - 3\mathbf{j} + 4\mathbf{k}$
   (g) $\mathbf{a} = -3\mathbf{i} + 6\mathbf{j} - \mathbf{k}, \mathbf{b} = 2\mathbf{i} - 2\mathbf{j} + 3\mathbf{k}$
   (h) $\mathbf{a} = 2\mathbf{i} - 6\mathbf{j} + 3\mathbf{k}, \mathbf{b} = 2\mathbf{i} - \mathbf{j} + 2\mathbf{k}$
   (i) $\mathbf{a} = 2\mathbf{i} + \mathbf{j} - 3\mathbf{k}, \mathbf{b} = \mathbf{i} + 2\mathbf{j} + \mathbf{k}$
   (j) $\mathbf{a} = 5\mathbf{i} - 3\mathbf{j} + 7\mathbf{k}, \mathbf{b} = -3\mathbf{i} - 4\mathbf{j} + 2\mathbf{k}$

3 Find the value of λ for which the vectors $2\mathbf{i} - 3\mathbf{j} + \mathbf{k}$ and $3\mathbf{i} + 6\mathbf{j} + \lambda\mathbf{k}$ are perpendicular.   [E]

4 Given that $\mathbf{u} = 3\mathbf{i} + 2\mathbf{j}$ and $\mathbf{v} = 2\mathbf{i} + \lambda\mathbf{j}$, determine the value of λ such that:

(a) $\mathbf{u}$ and $\mathbf{v}$ are at right angles

(b) $\mathbf{u}$ and $\mathbf{v}$ are parallel

(c) the acute angle between $\mathbf{u}$ and $\mathbf{v}$ is $\dfrac{\pi}{4}$. [E]

5 The angle between $\mathbf{i} + \mathbf{j}$ and $\mathbf{i} + \mathbf{j} + p\mathbf{k}$ is $\dfrac{\pi}{4}$. Find the possible values of p. [E]

6 Find the angle, in degrees to 1 decimal place, which the vector $\mathbf{a}$ makes with the positive (i) x-axis (ii) y-axis (iii) z-axis, where $\mathbf{a} =$

(a) $5\mathbf{i} + 2\mathbf{j} + 4\mathbf{k}$ (b) $8\mathbf{i} + \mathbf{j} - 4\mathbf{k}$ (c) $2\mathbf{i} - 3\mathbf{j} + 4\mathbf{k}$

(d) $-\mathbf{i} - \mathbf{j} - 3\mathbf{k}$ (e) $4\mathbf{i} - 2\mathbf{j} - \mathbf{k}$

7 Find a vector which is perpendicular to both $\mathbf{a}$ and $\mathbf{b}$ where:

(a) $\mathbf{a} = \mathbf{i} + 3\mathbf{j} - \mathbf{k}$, $\mathbf{b} = 3\mathbf{i} - \mathbf{j} - \mathbf{k}$

(b) $\mathbf{a} = 2\mathbf{i} - 3\mathbf{j} - 5\mathbf{k}$, $\mathbf{b} = \mathbf{i} - 2\mathbf{j} - 3\mathbf{k}$

(c) $\mathbf{a} = 2\mathbf{i} + 10\mathbf{j} + 3\mathbf{k}$, $\mathbf{b} = \mathbf{i} + 6\mathbf{j} + 2\mathbf{k}$

(d) $\mathbf{a} = 3\mathbf{i} + 2\mathbf{j} + \mathbf{k}$, $\mathbf{b} = -3\mathbf{i} + 4\mathbf{j} + 3\mathbf{k}$

(e) $\mathbf{a} = 2\mathbf{i} - 3\mathbf{j} - 4\mathbf{k}$, $\mathbf{b} = 4\mathbf{i} - 3\mathbf{j} + \mathbf{k}$

8 Find the values of λ for which the vectors $2\lambda\mathbf{i} + \lambda\mathbf{j} - 4\mathbf{k}$ and $\lambda\mathbf{i} - 2\mathbf{j} + \mathbf{k}$ are perpendicular.

9 Find the projection of $\mathbf{a}$ in the direction of $\mathbf{b}$ where:

(a) $\mathbf{a} = \mathbf{i} - 2\mathbf{j} + \mathbf{k}$, $\mathbf{b} = 4\mathbf{i} - 4\mathbf{j} + 7\mathbf{k}$

(b) $\mathbf{a} = 2\mathbf{i} - 3\mathbf{j} + 6\mathbf{k}$, $\mathbf{b} = \mathbf{i} + 2\mathbf{j} + 2\mathbf{k}$

(c) $\mathbf{a} = \mathbf{i} - \mathbf{j} + 2\mathbf{k}$, $\mathbf{b} = 2\mathbf{i} + 3\mathbf{j} - \mathbf{k}$

(d) $\mathbf{a} = -\mathbf{i} - 3\mathbf{j} + 5\mathbf{k}$, $\mathbf{b} = 2\mathbf{i} - 5\mathbf{j} + 4\mathbf{k}$

(e) $\mathbf{a} = 3\mathbf{i} - 5\mathbf{j} - 2\mathbf{k}$, $\mathbf{b} = 5\mathbf{i} + 4\mathbf{j} - \mathbf{k}$

10 The vectors $\mathbf{F}$ and $\mathbf{u}$ are $\mathbf{i} - 2\mathbf{j} + 3\mathbf{k}$ and $8\mathbf{i} + 9\mathbf{j} + 12\mathbf{k}$ respectively. Find the projection of $\mathbf{F}$ in the direction of $\mathbf{u}$.

11 Prove the cosine rule by vector methods.

12 Prove Pythagoras' theorem by vector methods.

13 Prove that the angle in a semi-circle is a right angle by vector methods.

14 Show that any one of $2\mathbf{i} - 2\mathbf{j} + \mathbf{k}$, $\mathbf{i} + 2\mathbf{j} + 2\mathbf{k}$ and $2\mathbf{i} + \mathbf{j} - 2\mathbf{k}$ is perpendicular to the other two.

15 Simplify $(\mathbf{a} + \mathbf{b}) \cdot \mathbf{c} - (\mathbf{a} - \mathbf{b}) \cdot \mathbf{c}$.

16 Points A and B have position vectors $\mathbf{a} = 2\mathbf{i} - 3\mathbf{j} + \mathbf{k}$, $\mathbf{b} = -4\mathbf{i} - \mathbf{j} + 3\mathbf{k}$ referred to an origin O. Calculate the sizes of the angles of $\triangle AOB$, giving your answers to the nearest tenth of a degree.

17 Given $A(2, 1, 7)$, $B(-3, 1, 4)$, $C(2, -1, 5)$, use a vector method to find the cosine of $\angle ABC$.

18 Given $A(-1, 2, -1)$, $B(2, 0, 5)$, $C(1, 5, -1)$, use a vector method to find the cosine of $\angle BCA$ and the area of $\triangle ABC$.

19 The vectors **a** and **b** are $2\mathbf{i} - \mathbf{j} + 3\mathbf{k}$ and $3\mathbf{i} + 2\mathbf{j} - 2\mathbf{k}$ respectively. Find the projection of **a** in the direction of **b**.

20 Find the angle between the vectors **a** and **b** given that $|\mathbf{a}| = 3$, $|\mathbf{b}| = 5$ and $|\mathbf{a} - \mathbf{b}| = 7$.   [E]

5.8 The vector equation of a straight line

Think of a line that passes through the point A and is parallel to the vector **b**. The point A has position vector **a** referred to the origin O. Let R be any other point on the line and let it have position vector **r**. Since the line is parallel to **b**, then:

$$\overrightarrow{AR} = \lambda\mathbf{b}, \text{ where } \lambda \text{ is a scalar}$$

However:

$$\overrightarrow{AR} = \mathbf{r} - \mathbf{a}$$

So: $\mathbf{r} - \mathbf{a} = \lambda\mathbf{b}$

or $\mathbf{r} = \mathbf{a} + \lambda\mathbf{b}$

This is a vector equation of the straight line. The vector **b** is in the same direction as the line and is sometimes called the **direction vector of the line**. The vector **a** is the position vector of a point on the line and λ is a scalar taking all real values.

Example 15
Find a vector equation of the line that passes through the point with position vector $2\mathbf{i} + \mathbf{j} - \mathbf{k}$ and is parallel to the vector $-5\mathbf{i} - 2\mathbf{j} - \mathbf{k}$.

The equation of the line is

$$\mathbf{r} = 2\mathbf{i} + \mathbf{j} - \mathbf{k} + \lambda(-5\mathbf{i} - 2\mathbf{j} - \mathbf{k})$$

This could also be rearranged as:

$$\mathbf{r} = (2 - 5\lambda)\mathbf{i} + (1 - 2\lambda)\mathbf{j} - (1 + \lambda)\mathbf{k}$$

Example 16

Find a vector equation of the line that passes through the points A and B with position vectors $\mathbf{a} = 2\mathbf{i} - 2\mathbf{j} + 3\mathbf{k}$ and $\mathbf{b} = -4\mathbf{i} + 5\mathbf{j} - \mathbf{k}$ respectively.

Since the line passes through the points A and B a direction vector for the line is $\overrightarrow{AB}$. (Notice that $\overrightarrow{BA}$ is also a direction vector for the line.) Then:

$$\overrightarrow{AB} = \mathbf{b} - \mathbf{a} = -4\mathbf{i} + 5\mathbf{j} - \mathbf{k} - (2\mathbf{i} - 2\mathbf{j} + 3\mathbf{k})$$
$$= -6\mathbf{i} + 7\mathbf{j} - 4\mathbf{k}$$

So a vector equation of the line is:

$$\mathbf{r} = 2\mathbf{i} - 2\mathbf{j} + 3\mathbf{k} + \lambda(-6\mathbf{i} + 7\mathbf{j} - 4\mathbf{k})$$

Notice that since B lies on the line, another form of the vector equation of the line is:

$$\mathbf{r} = -4\mathbf{i} + 5\mathbf{j} - \mathbf{k} + \mu(-6\mathbf{i} + 7\mathbf{j} - 4\mathbf{k})$$

Example 17

Find a vector equation of the line that passes through the points $A(1, 2, 1)$ and $B(2, -1, -1)$.

Show that the point $C(0, 5, 3)$ also lies on the line.

A direction vector for the line is:

$$\overrightarrow{AB} = 2\mathbf{i} - \mathbf{j} - \mathbf{k} - (\mathbf{i} + 2\mathbf{j} + \mathbf{k}) = \mathbf{i} - 3\mathbf{j} - 2\mathbf{k}$$

So an equation of the line is:

$$\mathbf{r} = \mathbf{i} + 2\mathbf{j} + \mathbf{k} + \lambda(\mathbf{i} - 3\mathbf{j} - 2\mathbf{k})$$

If $\lambda = -1$, then:

$$\mathbf{r} = 0\mathbf{i} + 5\mathbf{j} + 3\mathbf{k}$$

So the line passes through the point $(0, 5, 3)$.

Example 18

Show that the lines with equations

$$\mathbf{r} = 7\mathbf{i} - 3\mathbf{j} + 3\mathbf{k} + \lambda(3\mathbf{i} - 2\mathbf{j} + \mathbf{k})$$

and
$$\mathbf{r} = 7\mathbf{i} - 2\mathbf{j} + 4\mathbf{k} + \mu(-2\mathbf{i} + \mathbf{j} - \mathbf{k})$$

intersect and find the position vector of their point of intersection.

You can rewrite the first equation as:

$$\mathbf{r} = (7 + 3\lambda)\mathbf{i} + (-3 - 2\lambda)\mathbf{j} + (3 + \lambda)\mathbf{k}$$

and the second as:

$$\mathbf{r} = (7 - 2\mu)\mathbf{i} + (-2 + \mu)\mathbf{j} + (4 - \mu)\mathbf{k}$$

If these intersect then:

$$(7 + 3\lambda)\mathbf{i} + (-3 - 2\lambda)\mathbf{j} + (3 + \lambda)\mathbf{k} = (7 - 2\mu)\mathbf{i} + (-2 + \mu)\mathbf{j} + (4 - \mu)\mathbf{k}$$

Since **i**, **j** and **k** are non-parallel,

$$7 + 3\lambda = 7 - 2\mu \tag{1}$$
$$-3 - 2\lambda = -2 + \mu \tag{2}$$
$$3 + \lambda = 4 - \mu \tag{3}$$

From (1): $\qquad 3\lambda = -2\mu$ or $\mu = -\frac{3}{2}\lambda$

Substituting in (2): $\qquad -3 - 2\lambda = -2 - \frac{3}{2}\lambda$

$$\Rightarrow \quad -1 = \frac{1}{2}\lambda$$

$$\lambda = -2 \text{ so } \mu = 3$$

Check in (3): $\qquad 3 + \lambda = 3 - 2 = 1$

$$4 - \mu = 4 - 3 = 1$$

Notice that it is essential to check in the third equation, because if the values of λ and μ do not satisfy all *three* equations then it means that the lines do *not* intersect.

Using $\lambda = -2$ in the equation

$$\mathbf{r} = 7\mathbf{i} - 3\mathbf{j} + 3\mathbf{k} + \lambda(3\mathbf{i} - 2\mathbf{j} + \mathbf{k})$$

gives: $\qquad\qquad\qquad \mathbf{r} = \mathbf{i} + \mathbf{j} + \mathbf{k}$

where **r** is the position vector of the point of intersection of the two lines, as required.

Notice that using $\mu = 3$ in

$$\mathbf{r} = 7\mathbf{i} - 2\mathbf{j} + 4\mathbf{k} + \mu(-2\mathbf{i} + \mathbf{j} - \mathbf{k})$$

also gives $\mathbf{r} = \mathbf{i} + \mathbf{j} + \mathbf{k}$ and this in itself could be a sufficient check, since you have shown $\mathbf{i} + \mathbf{j} + \mathbf{k}$ to be on both lines.

5.9 Deriving cartesian equations of a straight line from a vector equation

From the vector equation of a straight line you have

$$\mathbf{r} = \mathbf{a} + \lambda\mathbf{b}$$

If $\qquad\qquad\qquad \mathbf{r} = x\mathbf{i} + y\mathbf{j} + z\mathbf{k}$

$$\mathbf{a} = a_1\mathbf{i} + a_2\mathbf{j} + a_3\mathbf{k}$$

and $\qquad\qquad\quad \mathbf{b} = b_1\mathbf{i} + b_2\mathbf{j} + b_3\mathbf{k}$

then: $\quad x\mathbf{i} + y\mathbf{j} + z\mathbf{k} = a_1\mathbf{i} + a_2\mathbf{j} + a_3\mathbf{k} + \lambda(b_1\mathbf{i} + b_2\mathbf{j} + b_3\mathbf{k})$

So: $\quad x\mathbf{i} + y\mathbf{j} + z\mathbf{k} = (a_1 + \lambda b_1)\mathbf{i} + (a_2 + \lambda b_2)\mathbf{j} + (a_3 + \lambda b_3)\mathbf{k}$

Since **i**, **j**, **k** are not parallel:

$$x = a_1 + \lambda b_1 \Rightarrow \lambda = \frac{x - a_1}{b_1}$$

$$y = a_2 + \lambda b_2 \Rightarrow \lambda = \frac{y - a_2}{b_2}$$

$$z = a_3 + \lambda b_3 \Rightarrow \lambda = \frac{z - a_3}{b_3}$$

Putting these values of λ together gives

$$\frac{x - a_1}{b_1} = \frac{y - a_2}{b_2} = \frac{z - a_3}{b_3}$$

which are the equations of a straight line in cartesian form.

■ So the cartesian equations of the line that passes through the point (a_1, a_2, a_3) and has direction vector $b_1\mathbf{i} + b_2\mathbf{j} + b_3\mathbf{k}$ are

$$\frac{x - a_1}{b_1} = \frac{y - a_2}{b_2} = \frac{z - a_3}{b_3}$$

Example 19

Find cartesian equations for the line that is parallel to the vector $2\mathbf{i} - 5\mathbf{j} + 4\mathbf{k}$ and passes through the point A with position vector $-\mathbf{i} + 7\mathbf{j} - 2\mathbf{k}$.

A vector equation of the line is:

$$\mathbf{r} = -\mathbf{i} + 7\mathbf{j} - 2\mathbf{k} + \lambda(2\mathbf{i} - 5\mathbf{j} + 4\mathbf{k})$$

or $\qquad \mathbf{r} = (-1 + 2\lambda)\mathbf{i} + (7 - 5\lambda)\mathbf{j} + (-2 + 4\lambda)\mathbf{k}$

So: $\qquad x = -1 + 2\lambda, \; y = 7 - 5\lambda, \; z = -2 + 4\lambda$

The cartesian equations are therefore:

$$\frac{x + 1}{2} = \frac{y - 7}{-5} = \frac{z + 2}{4}$$

Example 20

Find a vector equation of the line which has cartesian equations

$$\frac{x - 1}{5} = \frac{y + 1}{2} = \frac{z}{-5}$$

Let $\lambda = \dfrac{x - 1}{5} = \dfrac{y + 1}{2} = \dfrac{z}{-5}$

Then $x = 5\lambda + 1, \; y = 2\lambda - 1, \; z = -5\lambda$

So a vector equation is

$$\mathbf{r} = \mathbf{i} - \mathbf{j} + \lambda(5\mathbf{i} + 2\mathbf{j} - 5\mathbf{k})$$

Example 21

Find cartesian equations for the line that passes through the points $(3, 1, -2)$ and $(4, 3, 4)$.

A direction vector of the line is:

$$(4 - 3)\mathbf{i} + (3 - 1)\mathbf{j} + (4 + 2)\mathbf{k} = \mathbf{i} + 2\mathbf{j} + 6\mathbf{k}$$

So a vector equation of the line is:

$$\mathbf{r} = 3\mathbf{i} + \mathbf{j} - 2\mathbf{k} + \lambda(\mathbf{i} + 2\mathbf{j} + 6\mathbf{k})$$

Thus: $\qquad x = 3 + \lambda,\ y = 1 + 2\lambda,\ z = -2 + 6\lambda$

and cartesian equations of the line are

$$\frac{x - 3}{1} = \frac{y - 1}{2} = \frac{z + 2}{6}$$

■ **In general, a cartesian equation of the line passing through (x_1, y_1, z_1) and (x_2, y_2, z_2) is:**

$$\frac{x - x_1}{x_2 - x_1} = \frac{y - y_1}{y_2 - y_1} = \frac{z - z_1}{z_2 - z_1}$$

Example 22

The cartesian equations of two straight lines L_1 and L_2 are

$$L_1: \quad \frac{x - 1}{5} = \frac{y + 2}{4} = \frac{z}{6}$$

$$L_2: \quad \frac{x + 3}{2} = \frac{y - 5}{7} = \frac{z + 1}{-1}$$

Find the acute angle between the two lines.

For L_1: $\qquad x = 1 + 5\lambda,\ y = -2 + 4\lambda,\ z = 6\lambda$

So a vector equation of L_1 is

$$\mathbf{r} = \mathbf{i} - 2\mathbf{j} + \lambda(5\mathbf{i} + 4\mathbf{j} + 6\mathbf{k})$$

For L_2: $\qquad x = -3 + 2\mu,\ y = 5 + 7\mu,\ z = -1 - \mu$

So a vector equation of L_2 is

$$\mathbf{r} = -3\mathbf{i} + 5\mathbf{j} - \mathbf{k} + \mu(2\mathbf{i} + 7\mathbf{j} - \mathbf{k})$$

The angle between the lines is the angle between their direction vectors. The scalar product of the direction vectors is:

$$(5\mathbf{i} + 4\mathbf{j} + 6\mathbf{k}) \cdot (2\mathbf{i} + 7\mathbf{j} - \mathbf{k}) = 10 + 28 - 6 = 32$$

Also: $\qquad |5\mathbf{i} + 4\mathbf{j} + 6\mathbf{k}| = \sqrt{(25 + 16 + 36)} = \sqrt{77}$

and: $\qquad |2\mathbf{i} + 7\mathbf{j} - \mathbf{k}| = \sqrt{(4 + 49 + 1)} = \sqrt{54}$

So: $\qquad\qquad \sqrt{77}\ \sqrt{54} \cos\theta = 32$

$$\cos\theta = \frac{32}{\sqrt{77}\ \sqrt{54}} = 0.4963$$

So: $\qquad\qquad\qquad \theta = 60.2°$

Exercise 5E

1 Find a vector equation of the line that passes through the point
with position vector **a** and is parallel to the vector **b** where:
(a) $\mathbf{a} = -\mathbf{i} + \mathbf{j} + \mathbf{k}, \mathbf{b} = 2\mathbf{i} - 3\mathbf{j} + \mathbf{k}$
(b) $\mathbf{a} = 2\mathbf{i} - 7\mathbf{j} - 11\mathbf{k}, \mathbf{b} = -3\mathbf{i} + 9\mathbf{j} + 17\mathbf{k}$
(c) $\mathbf{a} = -3\mathbf{i} + 9\mathbf{j} - 15\mathbf{k}, \mathbf{b} = 2\mathbf{i} + \mathbf{j}$
(d) $\mathbf{a} = 3\mathbf{i} - \mathbf{k}, \mathbf{b} = -\mathbf{i} - 5\mathbf{j} + 2\mathbf{k}$
(e) $\mathbf{a} = 5\mathbf{i} - \mathbf{j} + 4\mathbf{k}, \mathbf{b} = \mathbf{i} - 2\mathbf{j} + \mathbf{k}$

2 Find a vector equation of the line that passes through the
point $(-1, 7, 4)$ and is parallel to the y-axis.

3 Find a vector equation of the line which passes through the
point $(2, -1, 6)$ and is perpendicular to the yz-plane.

4 Find vector equations of the lines passing through the
following points:
(a) $(1, -1, 2)$ and $(-5, 2, 2)$ (b) $(2, -1, 7)$ and $(7, 0, -3)$
(c) $(-3, -1, 2)$ and $(8, 1, -3)$ (d) $(2, -6, 4)$ and $(-1, -2, -3)$
(e) $(3, 1, 4)$ and $(-1, 2, 5)$

5 Show that the point $(11.5, 20.5, -28)$ lies on the line with
equation

$$\mathbf{r} = \mathbf{i} - 4\mathbf{j} + 7\mathbf{k} + \lambda(-3\mathbf{i} - 7\mathbf{j} + 10\mathbf{k})$$

and state the value of λ at this point.

6 Find cartesian equations of the lines with vector equations:
(a) $\mathbf{r} = -3\mathbf{i} + 2\mathbf{j} - 5\mathbf{k} + \lambda(2\mathbf{i} - 4\mathbf{j} + \mathbf{k})$
(b) $\mathbf{r} = \mathbf{i} - 3\mathbf{j} + 2\mathbf{k} + \lambda(-\mathbf{i} + \mathbf{j} - 2\mathbf{k})$
(c) $\mathbf{r} = -\mathbf{i} + 3\mathbf{j} + 4\mathbf{k} + \lambda(-6\mathbf{i} + 4\mathbf{j} - 9\mathbf{k})$
(d) $\mathbf{r} = 2\mathbf{i} - 3\mathbf{j} + 4\mathbf{k} + \lambda(-2\mathbf{i} + 6\mathbf{j} - \mathbf{k})$
(e) $\mathbf{r} = -3\mathbf{i} + 4\mathbf{j} - \mathbf{k} + \lambda(2\mathbf{i} - 3\mathbf{j} - \mathbf{k})$

7 Find a vector equation of the line with cartesian equations:
(a) $\dfrac{x-1}{2} = \dfrac{y+1}{4} = \dfrac{z-1}{3}$ (b) $\dfrac{x-1}{3} = \dfrac{y+1}{-4} = \dfrac{z-3}{-1}$

(c) $\dfrac{1-x}{2} = \dfrac{y+2}{4} = \dfrac{z+3}{5}$ (d) $\dfrac{1-x}{3} = \dfrac{4-y}{2} = \dfrac{z+3}{-1}$

(e) $\dfrac{2x-1}{4} = \dfrac{2-y}{3} = \dfrac{1-3z}{2}$

8 The lines with vector equations

$$\mathbf{r} = \mathbf{i} + 3\mathbf{j} - 2\mathbf{k} + \lambda(4\mathbf{i} + \mathbf{k})$$

and $$\mathbf{r} = 5\mathbf{i} + 3\mathbf{j} + 8\mathbf{k} + \mu(-\mathbf{i} + 2\mathbf{k})$$

intersect. Find the position vector of the point of intersection.

9 Find the coordinates of the point of intersection of the lines L_1 and L_2 with cartesian equations:

$$L_1: \quad x - 6 = \frac{-2 - y}{5} = \frac{z - 8}{7}$$

$$L_2: \quad \frac{x - 1}{3} = \frac{y - 3}{5} = \frac{2 - z}{8}$$

10 Determine whether the lines with the given equations intersect. If they do, state the coordinates of the point of intersection.

(a) $\dfrac{x}{4} = \dfrac{2 - y}{2} = \dfrac{z + 1}{3}$ and $4 - x = \dfrac{y - 1}{3} = \dfrac{z + 2}{4}$

(b) $\mathbf{r} = -3\mathbf{j} + 5\mathbf{k} + \lambda(2\mathbf{i} + 4\mathbf{j} + 2\mathbf{k})$ and
$\mathbf{r} = \mathbf{i} + 8\mathbf{j} + 3\mathbf{k} + \mu(-\mathbf{i} + \mathbf{j} - 2\mathbf{k})$

(c) $\mathbf{r} = 2\mathbf{i} - \mathbf{k} + \lambda(3\mathbf{i} + \frac{1}{2}\mathbf{j} + 4\mathbf{k})$ and
$\mathbf{r} = -4\mathbf{i} + 5\mathbf{j} + \frac{5}{3}\mathbf{k} + \mu(3\mathbf{i} + \mathbf{j} + \frac{4}{3}\mathbf{k})$

(d) $\dfrac{x - 14}{5} = y - 3 = \dfrac{z + 2}{2}$ and $\dfrac{x - 2}{3} = \dfrac{y + 5}{2} = 2 - z$

11 Find, in degrees to 1 decimal place, the acute angle between the lines with equations:

(a) $\mathbf{r} = \mathbf{i} + 2\mathbf{j} - \mathbf{k} + \lambda(-\mathbf{i} + 3\mathbf{j} - \mathbf{k})$ and
$\mathbf{r} = -\mathbf{i} + 2\mathbf{j} - \mathbf{k} + \mu(3\mathbf{i} - \mathbf{j} + 2\mathbf{k})$

(b) $\mathbf{r} = -\mathbf{i} + 3\mathbf{j} + 5\mathbf{k} + \lambda(-3\mathbf{i} + 4\mathbf{j} + \mathbf{k})$ and
$\mathbf{r} = 3\mathbf{i} - 2\mathbf{j} + \mathbf{k} + \mu(4\mathbf{i} - 2\mathbf{j} + 6\mathbf{k})$

(c) $\mathbf{r} = \mathbf{i} - 2\mathbf{k} + \lambda(-2\mathbf{i} + 3\mathbf{j} - \mathbf{k})$ and $\mathbf{r} = 2\mathbf{j} - 4\mathbf{k} + \mu(3\mathbf{i} - \mathbf{j} + 6\mathbf{k})$

(d) $\dfrac{x - 1}{2} = \dfrac{y + 2}{1} = 3 - z$ and $\dfrac{x + 1}{-2} = y - 3 = \dfrac{z - 7}{2}$

(e) $\dfrac{x - 2}{5} = \dfrac{3 - y}{3} = \dfrac{z + 1}{2}$ and $\dfrac{9 - x}{3} = \dfrac{y - 2}{5} = 2 - z$

12 Points A, B, C, D in a plane have position vectors $\mathbf{a} = 6\mathbf{i} + 8\mathbf{j}$, $\mathbf{b} = \frac{3}{2}\mathbf{a}$, $\mathbf{c} = 6\mathbf{i} + 3\mathbf{j}$, $\mathbf{d} = \frac{5}{3}\mathbf{c}$ respectively. Write down vector equations of the lines AD and BC and find the position vector of their point of intersection. **[E]**

13 State a vector equation of the line passing through the points A and B whose position vectors are $\mathbf{i} - \mathbf{j} + 3\mathbf{k}$ and $\mathbf{i} + 2\mathbf{j} + 2\mathbf{k}$ respectively. Determine the position vector of the point C which divides the line segment AB internally such that $AC = 2CB$. **[E]**

14 Show that the lines

$$\mathbf{r} = (-2\mathbf{i} + 5\mathbf{j} - 11\mathbf{k}) + \lambda(3\mathbf{i} + \mathbf{j} + 3\mathbf{k})$$
$$\mathbf{r} = 8\mathbf{i} + 9\mathbf{j} + \mu(4\mathbf{i} + 2\mathbf{j} + 5\mathbf{k})$$

intersect. Find the position vector of their common point. **[E]**

15 Find the point of intersection of the line through the points $(2, 0, 1)$ and $(-1, 3, 4)$ and the line through the points $(-1, 3, 0)$ and $(4, -2, 5)$. Calculate the acute angle between the two lines.   [E]

16 The point P lies on the line which is parallel to the vector $2\mathbf{i} + \mathbf{j} - \mathbf{k}$ and which passes through the point with position vector $\mathbf{i} + \mathbf{j} + 2\mathbf{k}$. The point Q lies on another line which is parallel to the vector $\mathbf{i} + \mathbf{j} - 2\mathbf{k}$ and which passes through the point with position vector $\mathbf{i} + \mathbf{j} + 4\mathbf{k}$. The line PQ is perpendicular to both these lines. Find a vector equation of the line PQ and the coordinates of the mid-point of PQ.   [E]

17 The position vectors of three points A, B and C with respect to a fixed origin O are

$$(2\mathbf{i} - 2\mathbf{j} + \mathbf{k}), \ (4\mathbf{i} + 2\mathbf{j} + \mathbf{k}) \text{ and } (\mathbf{i} + \mathbf{j} + 3\mathbf{k})$$

respectively.

(a) Write down unit vectors in the direction of the lines CA and CB and calculate the size of $\angle ACB$, in degrees, to 1 decimal place. The mid-point of AB is M.

(b) Find a vector equation of the straight line passing through C and M.

(c) Show that AB and CM are perpendicular.

(d) Find the position vector of the point N on the line CM such that $\angle ONC = 90°$.   [E]

18 Show that the lines L_1 and L_2 which have cartesian equations

$$L_1: \quad \frac{x - 9}{2} = -y - 4 = z - 5$$

$$L_2: \quad x - 2 = \frac{y + 8}{2} = \frac{12 - z}{3}$$

meet in a point P. Find the coordinates of P. The point A is on the line L_1 and has y-coordinate zero. Find the x and z coordinates of A.

19 The lines L_1 and L_2 intersect at the point B. Given that the equations of the lines are

$$L_1: \quad \mathbf{r} = (2 + 3t)\mathbf{i} + (3 + 4t)\mathbf{j} + (4 + 2t)\mathbf{k}$$
$$L_2: \quad \mathbf{r} = (1 + 2s)\mathbf{i} + (1 + 3s)\mathbf{j} + (4s - 2)\mathbf{k}$$

where t and s are scalar parameters, find

(a) the coordinates of B

(b) the acute angle, in degrees to 1 decimal place, between the lines.

20 Referred to an origin O, the points A, B, C and D are $(3, 1, -1)$, $(6, 7, 8)$, $(2, 5, 0)$ and $(0, 7, -2)$ respectively.

(a) Find vector equations for AB and CD.

(b) Show that AB and CD intersect at the point $(4, 3, 2)$.

(c) Calculate the size of $\angle ACD$ to the nearest degree.

SUMMARY OF KEY POINTS

1 A vector is a quantity which has both magnitude and direction in space.

2 A unit vector is a vector whose magnitude (modulus) is 1.

3 If **a** and **b** are parallel then $\mathbf{a} = \lambda\mathbf{b}$ for some scalar λ.

4 Vectors are added by the triangle law:

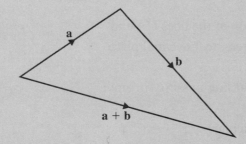

They can also be added or subtracted by the parallelogram law:

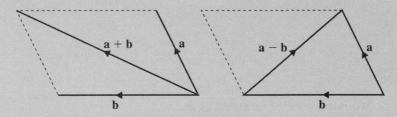

5 The vector $-\mathbf{a}$ has the same magnitude but opposite direction to **a**.

6 If **a** and **b** are non-parallel vectors and

$$\lambda\mathbf{a} + \mu\mathbf{b} = \alpha\mathbf{a} + \beta\mathbf{b}$$

where λ, μ, α, β are scalars, then

$$\lambda = \alpha \text{ and } \mu = \beta$$

7 If the points A and B have position vectors **a** and **b** respectively, then M, the mid-point of AB, has position vector $\frac{1}{2}(\mathbf{a} + \mathbf{b})$.

8 The distance between the points with coordinates (x_1, y_1, z_1) and (x_2, y_2, z_2) is

$$\sqrt{[(x_1 - x_2)^2 + (y_1 - y_2)^2 + (z_1 - z_2)^2]}.$$

9 The modulus of $\mathbf{a} = x\mathbf{i} + y\mathbf{j} + z\mathbf{k}$ is

$$|\mathbf{a}| = \sqrt{(x^2 + y^2 + z^2)}$$

10 If P has coordinates (x, y, z) then a unit vector in the direction $\overrightarrow{OP}$, where O is the origin, is

$$\frac{x\mathbf{i} + y\mathbf{j} + z\mathbf{k}}{\sqrt{(x^2 + y^2 + z^2)}}$$

11 The scalar (or dot) product of $\mathbf{a} = x_1\mathbf{i} + y_1\mathbf{j} + z_1\mathbf{k}$ and $\mathbf{b} = x_2\mathbf{i} + y_2\mathbf{j} + z_2\mathbf{k}$ is

$$\mathbf{a} \cdot \mathbf{b} = x_1 x_2 + y_1 y_2 + z_1 z_2 = |\mathbf{a}||\mathbf{b}| \cos\theta$$

where θ is the angle between $\mathbf{a}$ and $\mathbf{b}$.

12 The non-zero vectors $\mathbf{a}$ and $\mathbf{b}$ are perpendicular *if and only if* $\mathbf{a} \cdot \mathbf{b} = 0$.

13 The length of the projection of $\mathbf{a}$ on $\mathbf{b}$ is $\dfrac{\mathbf{a} \cdot \mathbf{b}}{|\mathbf{b}|}$

14 The vector equation of the straight line passing through the point with position vector $\mathbf{a}$ and parallel to the vector $\mathbf{b}$ is

$$\mathbf{r} = \mathbf{a} + \lambda\mathbf{b}$$

where λ is a scalar.

15 The straight line passing through the points A and B with position vectors $\mathbf{a}$ and $\mathbf{b}$ respectively has vector equation

$$\mathbf{r} = \mathbf{a} + \lambda(\mathbf{b} - \mathbf{a})$$

16 Cartesian equations of the straight line passing through the point (a_1, a_2, a_3) and parallel to the vector $b_1\mathbf{i} + b_2\mathbf{j} + b_3\mathbf{k}$ are

$$\frac{x - a_1}{b_1} = \frac{y - a_2}{b_2} = \frac{z - a_3}{b_3}$$

17 The straight line passing through (x_1, y_1, z_1) and (x_2, y_2, z_2) has cartesian equations

$$\frac{x - x_1}{x_2 - x_1} = \frac{y - y_1}{y_2 - y_1} = \frac{z - z_1}{z_2 - z_1}$$

Review exercise 2

1 Referred to an origin O, the points A and B have position
  vectors given by:
$$\overrightarrow{OA} = 5\mathbf{i} - 6\mathbf{j} - 2\mathbf{k} \text{ and } \overrightarrow{OB} = 2\mathbf{i} - 2\mathbf{j} + 10\mathbf{k}$$
  (a) Find $|\overrightarrow{AB}|$.
  (b) Calculate $\angle AOB$ to the nearest degree.

2 Express $\dfrac{1}{x(x+3)}$ in partial fractions.

  Hence show that
$$\int_1^3 \frac{1}{x(x+3)} \, dx = \tfrac{1}{3} \ln 2 \qquad \text{[E]}$$

3 Find

  (a) $\displaystyle\int x \cos x \, dx$

  (b) $\displaystyle\int \cos^2 y \, dy$

  Hence find the general solution of the differential equation
$$\frac{dy}{dx} = x \cos x \sec^2 y, \; 0 < y < \frac{\pi}{2} \qquad \text{[E]}$$

4 Sketch the curve defined with parameter ϕ by the equations
$$x = a \cos \phi, \; y = b \sin \phi$$
  where a and b are positive constants and $0 \leqslant \phi < 2\pi$.
  Given that P is the point at which $\phi = \dfrac{\pi}{4}$, obtain
  (a) an equation of the tangent to the curve at P,
  (b) an equation of the normal to the curve at P.
  Determine the area of the finite region bounded by the curve.

  [E]

5 Referred to the origin O, the points A and B have position vectors $(4\mathbf{i} - 11\mathbf{j} + 4\mathbf{k})$ and $(7\mathbf{i} + \mathbf{j} + 7\mathbf{k})$ respectively.

(a) Find a vector equation for the line passing through A and B in terms of a parameter t.

(b) Find the position vector of the point P on AB such that OP is perpendicular to AB. [E]

6 Express as the sum of partial fractions

$$\frac{2}{x(x+1)(x+2)}$$

Hence show that

$$\int_{2}^{4} \frac{2}{x(x+1)(x+2)}\,\mathrm{d}x = 3\ln 3 - 2\ln 5$$

7 The figure shows a sketch of the curve with parametric equations

$$x = \cos^2 t, \quad y = \tfrac{3}{2}\sin t, \quad -\frac{\pi}{2} \leqslant t \leqslant \frac{\pi}{2}$$

(a) Find the coordinates of the point where the curve crosses the x-axis.

(b) Find $\dfrac{\mathrm{d}x}{\mathrm{d}t}$

(c) Show that the area A of the finite region bounded by the curve and the y-axis is given by $A = 6\displaystyle\int_{0}^{\frac{\pi}{2}} \sin^2 t \cos t\,\mathrm{d}t.$

Using the substitution $u = \sin t$, or otherwise, evaluate A. [E]

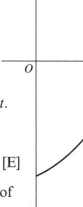

8 Given that $y > \tfrac{1}{2}$ and $0 < x < \tfrac{1}{4}\pi$, find the general solution of the differential equation

$$\left(\frac{\mathrm{d}y}{\mathrm{d}x}\right)\tan 2x = 2y - 1$$

Given further that $y = 1$ at $x = \tfrac{1}{8}\pi$, find the value of y at $x = \tfrac{1}{6}\pi$. [E]

9 The position vectors of the points A and B relative to an origin O are $5\mathbf{i} + 4\mathbf{j} + \mathbf{k}$, $-\mathbf{i} + \mathbf{j} - 2\mathbf{k}$ respectively. Find the position vector of the point P which lies on AB produced such that $AP = 2BP$. [E]

10 (a) Obtain the general solution of the differential equation

$$\frac{dy}{dx} = xy^2, \qquad y > 0$$

(b) Given also that $y = 1$ at $x = 1$ show that

$$y = \frac{2}{3 - x^2}, \qquad -\sqrt{3} < x < \sqrt{3}$$

is a particular solution of the differential equation.

The curve C has equation $y = \dfrac{2}{3 - x^2}$, $x \neq -\sqrt{3}$, $x \neq \sqrt{3}$.

(c) Write down the gradient of C at the point $(1, 1)$.

(d) Deduce that the line which is a tangent to C at the point $(1, 1)$ has equation $y = x$.

(e) Find the coordinates of the point where the line $y = x$ again meets the curve C. [E]

11 (a) Find $\displaystyle\int x \sin 2x \, dx$.

(b) Given that $y = 0$ at $x = \dfrac{\pi}{4}$, solve the differential equation

$$\frac{dy}{dx} = x \sin 2x \cos^2 y$$ [E]

12 Sketch the curve given parametrically by

$$x = t^2 - 2, \quad y = 2t, \quad t \in \mathbb{R}$$

indicating on your sketch where (a) $t = 0$ (b) $t > 0$ (c) $t < 0$.
Calculate the area of the finite region enclosed by the curve and the y-axis. [E]

13 With respect to an origin O, the position vectors of the points L, M and N are $(4\mathbf{i} + 7\mathbf{j} + 7\mathbf{k})$, $(\mathbf{i} + 3\mathbf{j} + 2\mathbf{k})$ and $(2\mathbf{i} + 4\mathbf{j} + 6\mathbf{k})$ respectively.

(a) Find the vectors $\overrightarrow{ML}$ and $\overrightarrow{MN}$.

(b) Prove that $\cos \angle LMN = \frac{9}{10}$. [E]

14 $$f(x) \equiv \frac{p - 2x}{x + q}, \quad x \in \mathbb{R}, \ x \neq -q,$$

where p and q are constants. The curve C, with equation $y = f(x)$, has an asymptote with equation $x = 2$ and passes through the point with coordinates $(3, 2)$.

(a) Write down the value of q.

(b) Show that $p = 8$.

(c) Write down an equation of the second asymptote to C.

(d) Using $p = 8$ and the value of q found in (a), sketch the graph of C showing clearly how C approaches the asymptotes and the coordinates of the points where C intersects the axes.

The finite region A is bounded by C, the x-axis and the line with equation $x = 3$.

(e) Show that the area of A can be written in the form $r + s\ln 2$, where r and s are integers to be found. [E]

15 A curve is given parametrically by

$$x = \sin t, \ y = \cos^3 t, \ -\pi < t \leqslant \pi$$

Show that

(a) $-1 \leqslant x \leqslant 1$ and $-1 \leqslant y \leqslant 1$

(b) $\dfrac{dy}{dx} = a\sin 2t$, where a is constant, and give the value of a.

Find the value of $\dfrac{dy}{dx}$ at $x = 0$ and show that the curve has points of inflexion where $t = -\dfrac{3\pi}{4}, \ -\dfrac{\pi}{4}, \dfrac{\pi}{4}$ and $\dfrac{3\pi}{4}$.

Sketch the curve. [E]

16
$$f(x) \equiv \frac{3x}{(x+2)(x-1)}$$

(a) Express $f(x)$ in partial fractions.

(b) Hence evaluate $\displaystyle\int_{-1}^{0} f(x)\,dx$. [E]

17 In the diagram, M is the mid-point of OA and $ON : NB = 3 : 1$. The lines AN and BM intersect at the point X. $\overrightarrow{OA} = \mathbf{a}$ and $\overrightarrow{OB} = \mathbf{b}$.

(a) Write down, in terms of $\mathbf{a}$ and $\mathbf{b}$, expressions for $\overrightarrow{AN}$ and $\overrightarrow{BM}$.

Given that $\overrightarrow{AX} = h\overrightarrow{AN}$, where h is a scalar,

(b) show that $\overrightarrow{OX} = (1 - h)\mathbf{a} + \frac{3}{4}h\mathbf{b}$.

Given that $\overrightarrow{BX} = k\overrightarrow{BM}$, where k is a scalar,

(c) find an expression for $\overrightarrow{OX}$ in terms of $\mathbf{a}$, $\mathbf{b}$ and k.

Using the expressions for $\overrightarrow{OX}$ found in (b) and (c),

(d) calculate the values of h and k.

(e) Write down the value of the ratio $AX : XN$. [E]

18 At time t, the rate of increase in the concentration x of micro-organisms in controlled surroundings is equal to k times the concentration, where k is a positive constant.

(a) Write down a differential equation in x, t and k.

(b) Given that $x = c$ at time $t = 0$, find x in terms of c, k and t.

(c) Sketch the graph of x against t for $t \geqslant 0$.

(d) Given further that $k = 10^{-2}$ and that t is measured in hours, find the time taken for the concentration to increase to $\dfrac{3c}{2}$ from when it was c. [E]

19
$$f(x) \equiv 6x^3 + Ax^2 + x - 2$$

where A is a constant.

Given that $x + 2$ is a factor of $f(x)$,

(a) find the value of A.

Using this value of A,

(b) express $\dfrac{1}{f(x)}$ in partial fractions,

(c) find the value of $\dfrac{d^2}{dx^2}\left(\dfrac{1}{f(x)}\right)$ at $x - 0$. [E]

20. The curve C is given by
$$x = 2t - 1, \quad y = 4 - t^2,$$

where t is a parameter.

The line with equation $x + 2y = 3$ meets C at the points P and Q.

(a) Determine the coordinates of P and Q.

The tangent at the point T to the curve C is parallel to PQ.

(b) Find the coordinates of T.

The finite region R is bounded by C and the line PQ.

(c) Draw a sketch to show the curve C, the line PQ and the region R.

(d) Using integration, find the area of R. [E]

21 $ORST$ is a parallelogram, U is the mid-point of RS and V is the mid-point of ST. Relative to the origin O, $\mathbf{r}$, $\mathbf{s}$, $\mathbf{t}$, $\mathbf{u}$ and $\mathbf{v}$ are the position vectors of R, S, T, U and V respectively.

(a) Express $\mathbf{s}$ in terms of $\mathbf{r}$ and $\mathbf{t}$.

(b) Express $\mathbf{v}$ in terms of $\mathbf{s}$ and $\mathbf{t}$.

(c) Hence, or otherwise, show that
$$4(\mathbf{u} + \mathbf{v}) = 3(\mathbf{r} + \mathbf{s} + \mathbf{t})$$ [E]

22 The figure shows part of the curve with equation

$$x^2 = y(4 - y)^2$$

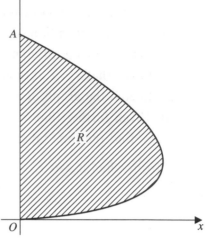

(a) Find the coordinates of the point A where the curve meets the y-axis.

The finite region R shaded in the figure is bounded by the curve with equation $x^2 = y(4 - y)^2$ and the line OA, where O is the origin.

(b) Calculate the area of R.

The finite region R is rotated through 2π about the y-axis.

(c) Calculate the volume of the solid formed, leaving your answer in terms of π. [E]

23 The normal at every point on a given curve passes through the point $(2, 3)$. Prove that the equation of the curve satisfies the differential equation

$$(x - 2) + \frac{\mathrm{d}y}{\mathrm{d}x}\,(y - 3) = 0$$

Hence find a cartesian equation of the family of curves which satisfy the condition above.

Find the equation of the particular curve which passes through the point $(4, 2)$. [E]

24 (a) Using the substitution $u^2 = x - 1$, or otherwise, prove that

$$\int_2^5 \frac{1}{1 + \sqrt{(x - 1)}}\,\mathrm{d}x = 2 - \ln(\tfrac{9}{4})$$

(b) Express

$$\frac{x^2 + x + 1}{(x + 1)(x^2 + 1)}$$

in partial fractions and hence prove that

$$\int_0^1 \frac{x^2 + x + 1}{(x + 1)(x^2 + 1)}\,\mathrm{d}x = \tfrac{3}{4}\ln 2 + \frac{\pi}{8} \qquad [E]$$

25 Referred to a fixed origin O, the points A, B and C are given by

$$\overrightarrow{OA} = 9\mathbf{i}, \ \overrightarrow{OB} = 3\mathbf{j} \text{ and } \overrightarrow{OC} = 2\mathbf{k}$$

(a) Obtain $\overrightarrow{CA}$ and $\overrightarrow{CB}$.

(b) Calculate the size of $\angle ACB$, giving your answer to the nearest tenth of a degree. [E]

26 Newton's law of cooling states that the rate at which a body cools is directly proportional to the excess temperature of the body over the temperature of its immediate surroundings. Given that at time t minutes a body has a temperature $T\,°C$ and its surroundings a constant temperature $\theta\,°C$, form a differential equation in terms of T, θ, t, and the constant of proportionality, k, $k > 0$.

Integrate this equation to show that

$$\ln(T - \theta) = -kt + c$$

where c is a constant.

Hence show that

$$T = \theta + A\mathrm{e}^{-kt}$$

where $\ln A = c$.

At 2.23 p.m., the water in a kettle boils at $100\,°C$ in a room of constant temperature $21\,°C$. After 10 minutes, the temperature of the water in the kettle is $84\,°C$. Use this information to find the values of k and A, giving your answers to 2 significant figures. Hence find the time, to the nearest minute, when the temperature of the water in the kettle will be $70\,°C$.   [E]

27 It is given that

$$\mathrm{f}(x) \equiv \frac{3x + 1}{(x + 1)(x + 2)(x + 3)}$$

(a) Express $\mathrm{f}(x)$ as the sum of three partial fractions.

(b) Find $\displaystyle\int \mathrm{f}(x)\,\mathrm{d}x$.

(c) Hence show that the area of the finite region bounded by the curve with equation $y = \mathrm{f}(x)$ and the lines with equations $y = 0$, $x = 0$ and $x = 1$ is $\ln 2$.   [E]

28 Given that $\mathrm{e}^{2x+y}\,\dfrac{\mathrm{d}y}{\mathrm{d}x} = x$ and that $y = 0$ at $x = 0$, express e^y in terms of x.   [E]

29 The point A has coordinates $(7, -1, 3)$ and the point B has coordinates $(10, -2, 2)$. The line l has vector equation

$\mathbf{r} = \mathbf{i} + \mathbf{j} + \mathbf{k} + \lambda(3\mathbf{i} - \mathbf{j} + \mathbf{k})$, where λ is a real parameter.

(a) Show that the point A lies on the line l.

(b) Find the length of AB.

(c) Find the size of the acute angle between the line l and the line segment AB, giving your answer to the nearest degree.

(d) Hence, or otherwise, calculate the perpendicular distance from B to the line l, giving your answer to 2 significant figures. [E]

30 The finite region bounded by the curve with equation $y = \sin 2x$ and that part of the x-axis for which $0 \leqslant x \leqslant \frac{1}{2}\pi$, is rotated completely about the x-axis. Find, in terms of π, the volume of the solid of revolution generated. [E]

31 A curve C has parametric equations

$$x = 3t^2 - p, \quad y = 3t^3$$

where p is a non-negative constant and t is a real parameter.

(a) State, in terms of p, the range of values of x.

(b) Express $\dfrac{dy}{dx}$ in terms of t.

(c) Find the gradient of C at the point $(-p, 0)$.

(d) Sketch C.

When $p = 0$, the tangent to C, at the point where $t = \dfrac{2\sqrt{2}}{3}$,

meets C again at the point Q.

(e) Find an equation for this tangent.

(f) Verify that at Q, $t = -\dfrac{\sqrt{2}}{3}$.

(g) Deduce that this tangent is a normal to C at Q. [E]

32 Find the value of each of the constants A, B and C for which

$$\frac{7x^2 - 12x - 1}{(2x+1)(x-1)^2} \equiv \frac{A}{2x+1} + \frac{B}{x-1} + \frac{C}{(x-1)^2}$$

Hence evaluate $\displaystyle\int_2^5 \frac{7x^2 - 12x - 1}{(2x+1)(x-1)^2}\,dx$ [E]

33 Referred to a fixed origin O, the points A and B have position vectors $(5\mathbf{i} - \mathbf{j} - \mathbf{k})$ and $(\mathbf{i} - 5\mathbf{j} + 7\mathbf{k})$ respectively.

(a) Find an equation of the line AB.

(b) Show that the point C with position vector $4\mathbf{i} - 2\mathbf{j} + \mathbf{k}$ lies on AB.

(c) Show that OC is perpendicular to AB.

(d) Find the position vector of the point D, where $D \not\equiv A$, on AB such that $|\overrightarrow{OD}| = |\overrightarrow{OA}|$. [E]

34 (a) Given that $3y = 3\tan x + \tan^3 x$, show that

$$\frac{dy}{dx} = \sec^4 x$$

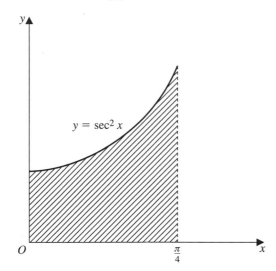

The shaded region is bounded by the x-axis, the y-axis, the

line $x = \dfrac{\pi}{4}$ and the curve with equation $y = \sec^2 x$. The region

is rotated through 2π about the x-axis to form a solid of
revolution.

(b) By integration, determine the volume of this solid, giving
your answer in terms of π.

35 Find the general solution of the differential equation

$$e^x \frac{dy}{dx} + y^2 = xy^2$$

Sketch, for positive values of x, the integral curve for which
$y = e$ at $x = 1$, showing all asymptotes and turning points. [E]

36 Find (a) $\displaystyle\int \frac{e^x}{e^x + 6}\,dx$ (b) $\displaystyle\int \cos^5 x \sin x\,dx$ (c) $\displaystyle\int \frac{x^2}{x+1}\,dx$.

37 Referred to an origin O, the points A and B have position
vectors given by

$$\overrightarrow{OA} = 7\mathbf{i} + 3\mathbf{j} + 8\mathbf{k}$$
$$\overrightarrow{OB} = 5\mathbf{i} + 4\mathbf{j} + 6\mathbf{k}$$

(a) Show that the point P with position vector given by

$$\overrightarrow{OP} = (5 - 2\lambda)\mathbf{i} + (4 + \lambda)\mathbf{j} + (6 - 2\lambda)\mathbf{k}$$

where λ is a parameter, lies on the straight line L passing
through the points A and B.

(b) Find the value of λ for which OP is perpendicular to L.

With centre O and radius OA, a circle is drawn to cut the line L at the points A and C.

(c) Determine the position vector of C.   [E]

38 Using the identity $\tan x \equiv \dfrac{\sin x}{\cos x}$, $-\dfrac{\pi}{2} < x < \dfrac{\pi}{2}$, show that

$$\frac{\mathrm{d}}{\mathrm{d}x}(\tan x) = \sec^2 x$$

Use integration by parts to find

$$\int x \sec^2 x \,\mathrm{d}x \qquad\qquad \text{[E]}$$

39 A curve is defined by the equations

$$x = t^2 - 1, \quad y = t^3 - t$$

where t is a parameter.

Sketch the curve for all real values of t.

Find the area of the region enclosed by the loop of the curve.

     [E]

40 Sketch the curve with equation $y = (2x - 1)^2(x + 1)$, showing the coordinates of

(a) the points where it meets the axes

(b) the turning points

(c) the point of inflexion.

By using your sketch, or otherwise, sketch the graph of

$$y = \frac{1}{(2x - 1)^2(x + 1)}$$

Show clearly the coordinates of any turning points.   [E]

41 Vectors **r** and **s** are given by

$$\mathbf{r} = \lambda\mathbf{i} + (2\lambda - 1)\mathbf{j} - \mathbf{k}$$
$$\mathbf{s} = (1 - \lambda)\mathbf{i} + 3\lambda\mathbf{j} + (4\lambda - 1)\mathbf{k}$$

where λ is a scalar.

(a) Find the values of λ for which **r** and **s** are perpendicular.

When $\lambda = 2$, **r** and **s** are the position vectors of the points A and B respectively, referred to an origin O.

(b) Find $\overrightarrow{AB}$.

(c) Use a scalar product to find the size of $\angle BAO$, giving your answer to the nearest degree.   [E]

42 (a) Evaluate

$$\int_0^{\frac{\pi}{4}} x \cos 2x \, \mathrm{d}x$$

(b) By using the substitution $u^2 = a^2 - x^2$, or otherwise, evaluate

$$\int_0^a x(a^2 - x^2)^{\frac{1}{2}} \, \mathrm{d}x \qquad \text{[E]}$$

43 Find the coordinates of the points of intersection of the curves with equations

$$y = \frac{x}{x+3} \text{ and } y = \frac{x}{x^2+1}$$

Sketch the curves on the same diagram, showing any asymptotes or turning points.

Show that the area of the finite region in the first quadrant enclosed by the two curves is

$$\tfrac{7}{2}\ln 5 - 3\ln 3 - 2 \qquad \text{[E]}$$

44 Solve the differential equation

$$\frac{\mathrm{d}y}{\mathrm{d}x} = x\mathrm{e}^{x+3y}$$

given that $y = 0$ at $x = 1$. \qquad [E]

45 Write down, in vector form, an equation of the line l which passes through $L(-3, 1, -7)$ and $M(5, 3, 5)$.

Find the position vector of the point P on the line for which OP is perpendicular to l, where O is the origin.

Hence find the shortest distance from O to the line l. \qquad [E]

46 Given that $\mathrm{f}(x) \equiv \mathrm{e}^{-x}(1 + x^2)$, find $\mathrm{g}(x)$ such that $\mathrm{f}'(x) \equiv \mathrm{e}^{-x}\mathrm{g}(x)$.

Hence, or otherwise, find the general solution of the differential equation $\mathrm{e}^x\mathrm{e}^{2y}\dfrac{\mathrm{d}y}{\mathrm{d}x} + (1 - x)^2 = 0$.

Express y in terms of x. \qquad [E]

47 Sketch the curve given by the parametric equations

$$x = 1 + t, \, y = 2 + \frac{1}{t} \, (t \neq 0)$$

Show that an equation of the normal to the curve at the point where $t = 2$ is $8x - 2y - 19 = 0$.

Find the value of t at the point where this normal meets the curve again. \qquad [E]

48

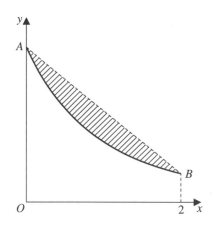

The figure shows a straight line joining the points A and B, together with the sketch-graph of a certain function. The function is known to be

either   (i) f: $x \mapsto x^2 - 3.6x + 4, \quad 0 \leqslant x \leqslant 2$

or   (ii) g: $x \mapsto \dfrac{4}{2x + 1}, \quad 0 \leqslant x \leqslant 2$

(a) Show that $f(0) = g(0)$ and that $f(2) = g(2)$.

(b) Calculate the area of the shaded region, giving your answer to two decimal places,

 (i) for the function f

(ii) for the function g.   [E]

49 The vectors **u** and **v** are given by

$$\mathbf{u} = 5\mathbf{i} - 4\mathbf{j} + s\mathbf{k}, \qquad \mathbf{v} = 2\mathbf{i} + t\mathbf{j} - 3\mathbf{k}$$

(a) Given that the vectors **u** and **v** are perpendicular, find a relation between the scalars s and t.

(b) Given instead that the vectors **u** and **v** are parallel, find the values of the scalars s and t.   [E]

50 The function f is defined on the domain $-e \leqslant x \leqslant e$ by

$$f(x) = \sqrt{x}, \qquad 0 \leqslant x \leqslant 1$$
$$f(x) = 1 - 2\ln x, \; 1 < x \leqslant e$$
$$\text{f is an odd function}$$

(a) Sketch the curve $y = f(x)$, marking the coordinates of the points where it crosses the x-axis.

(b) Calculate the area of the region in the first quadrant enclosed by the curve and the x-axis.   [E]

51 Sketch the curve given parametrically by the equations

$$x = 1 + t, \ y = 4 - t^2$$

Write on your sketch the coordinates of any points where the curve crosses the coordinate axes and the coordinates of the stationary point. Show that an equation of the normal to the curve at the point with parameter t is

$$x - 2ty = 2t^3 - 7t + 1$$

The normal to the curve at the point P where $t = 1$ cuts the curve again at the point Q. Determine the coordinates of Q.   [E]

52 Show that the substitution $y = vx$, where v is a function of x, transforms the differential equation

$$xy \frac{dy}{dx} = x^2 + y^2$$

into the differential equation

$$vx \frac{dv}{dx} = 1$$

Hence solve the original differential equation given that $y = 2$ at $x = 1$.   [E]

53 With respect to an origin O, the position vectors of the points L and M are $2\mathbf{i} - 3\mathbf{j} + 3\mathbf{k}$ and $5\mathbf{i} + \mathbf{j} + c\mathbf{k}$ respectively, where c is a constant. The point N is such that $OLMN$ is a rectangle.
 (a) Find the value of c.
 (b) Write down the position vector of N.
 (c) Find, in the form $\mathbf{r} = \mathbf{p} + t\mathbf{q}$, an equation of the line MN.
    [E]

54 (a) Use the substitution $u = \cos x$ to evaluate

$$\int_0^{\frac{\pi}{2}} \frac{\sin x}{3 + 5\cos x} \, dx$$

 (b) Given that $t = \tan \frac{x}{2}$, prove that $\cos x = \frac{1 - t^2}{1 + t^2}$ and

$$\frac{dx}{dt} = \frac{2}{1 + t^2}$$

Hence evaluate $\displaystyle\int_0^{\frac{\pi}{2}} \frac{1}{3 + 5\cos x} \, dx$.

55 A curve is given parametrically by the equations

$$x = c(1 + \cos t), \quad y = 2c\sin^2 t, \quad 0 \leqslant t \leqslant \pi$$

where c is a positive constant.

(a) Show that $\dfrac{dy}{dx} = -4\cos t$.

At the point P on the curve, $\cos t = \tfrac{3}{4}$.

(b) Show that the normal to the curve at P has equation
$$24y - 8x - 7c = 0$$

(c) Sketch the curve for $0 \leqslant t \leqslant \pi$.

The finite region R is bounded by the curve and the x-axis.

(d) Show that the area of R is given by the definite integral

$$2c^2 \int_0^\pi \sin^3 t\, dt$$

(e) Evaluate this integral to find the area of R in terms of c.

[E]

56 Solve the differential equation $x\dfrac{dy}{dx} = y + x^2 y$, given that

$y = 2e^{\frac{1}{2}}$ at $x = 1$. [E]

57 Show that the normal at the point $P\left(\dfrac{\pi}{4}, 1\right)$ to the curve with

equation $y = \tan x$ cuts the x-axis at the point $Q\left[\left(\dfrac{\pi}{4} + 2\right), 0\right]$.

The finite region A is bounded by the arc of the curve with equation $y = \tan x$ from the origin O to the point P, the line PQ and the line OQ. Find the area of A.

The region A is rotated through four right angles about the x-axis to generate the solid of revolution S. Find the volume of S. [E]

58 Two points A and B have position vectors $\mathbf{a}$ and $\mathbf{b}$ respectively. Show that the point which divides AB internally in the ratio $m : n$ has position vector

$$\frac{n\mathbf{a} + m\mathbf{b}}{n + m}$$

Three non-collinear points A, B and C have position vectors $\mathbf{a}$, $\mathbf{b}$ and $\mathbf{c}$ respectively. The point D divides AB internally in the ratio $2 : 1$. The point E divides BC internally in the ratio $2 : 1$. Show that DE produced meets AC produced at the point with position vector

$$\tfrac{4}{3}\mathbf{c} - \tfrac{1}{3}\mathbf{a}$$

[E]

59 The curve C has parametric equations

$$x = at, \quad y = \frac{a}{t}, \quad t \in \mathbb{R}, t \neq 0$$

where t is a parameter and a is a positive constant.

(a) Sketch C.

(b) Find $\dfrac{dy}{dx}$ in terms of t.

The point P on C has parameter $t = 2$.

(c) Show that an equation of the normal to C at P is

$$2y = 8x - 15a$$

This normal meets C again at the point Q.

(d) Find the value of t at Q. [E]

60 Express $\dfrac{x}{(1+x)(1+2x)}$ in the form $\dfrac{A}{1+x} + \dfrac{B}{1+2x}$, where

A and B are constants.

Given that $x > 0$, find the general solution of the differential equation

$$(1+x)(1+2x)\frac{dy}{dx} - xe^{2y} \qquad [E]$$

61 A line l_1 passes through the point A, with position vector $5\mathbf{i} + 3\mathbf{j}$, and the point B, with position vector $-2\mathbf{i} - 4\mathbf{j} + 7\mathbf{k}$.

(a) Write down an equation of the line l_1.

A second line l_2 has equation $\mathbf{r} = \mathbf{i} - 3\mathbf{j} - 4\mathbf{k} + \mu(\mathbf{i} + 2\mathbf{j} + 3\mathbf{k})$ where μ is a parameter.

(b) Show that l_1 and l_2 are perpendicular to each other.

(c) Show that the two lines meet, and find the position vector of the point of intersection.

The point C has position vector $2\mathbf{i} - \mathbf{j} - \mathbf{k}$.

(d) Show that C lies on l_2.

The point D is the image of C after reflection in the line l_1.

(e) Find the position vector of D. [E]

62 Evaluate the integrals

(a) $\displaystyle\int_0^{\frac{3}{4}} x\sqrt{(1+x^2)}\,dx$ (b) $\displaystyle\int_0^{\frac{\pi}{4}} \tan^2 x\,dx$

(c) $\displaystyle\int_0^{\pi} x\sin x\,dx$ (d) $\displaystyle\int_3^4 \frac{1}{x^2 - 3x + 2}\,dx$

Express your answer to (d) as a natural logarithm. [E]

63 The finite region R is bounded by the curve with equation $y = e^{2x}$, the coordinate axes and the line $x = 1$. The region R is rotated through 4 right angles about the x-axis. Find the volume generated, giving your answer to 2 significant figures. [E]

64 Sketch the curve given by the parametric equations

$$x = t^2 - 1, \quad y = 2t + 2, \quad t \in \mathbb{R}$$

The normal to this curve at the point P given by $t = 2$ meets the curve again at Q. Find the value of t at Q. [E]

65

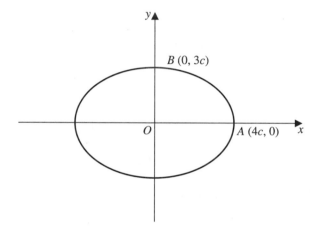

The figure shows a sketch of the curve given parametrically by the equations

$$x = 4c \cos t, \quad y = 3c \sin t, \quad -\pi < t \leqslant \pi$$

where c is a positive constant.

(a) Write down the value of t at the point $A(4c, 0)$ and at the point $B(0, 3c)$.

(b) By considering the integral $\int y \dfrac{dx}{dt} \, dt$ find, in terms of c,

the area of the region enclosed by the curve. [E]

66 (a) Find $\displaystyle\int \sin\theta \sin 2\theta \, d\theta$.

(b) Find the solution of the differential equation

$$(1 + e^{2y}) \frac{dy}{dx} = e^y \sin x \sin 2x$$

given that $y = 0$ at $x = \tfrac{1}{6}\pi$. [E]

67
$$f(x) \equiv \frac{x^2 + 6x + 7}{(x+2)(x+3)}, \quad x \in \mathbb{R}$$

Given that $f(x) \equiv A + \dfrac{B}{x+2} + \dfrac{C}{x+3}$,

(a) find the values of the constants A, B and C.

(b) Show that $\displaystyle\int_0^2 f(x)\,dx = 2 + \ln\left(\dfrac{25}{18}\right)$.　　　　　　　　[E]

68 (a) By using the substitution $u^2 = x - 1$, or otherwise, find

$$\int \frac{x+1}{\sqrt{(x-1)}}\,dx$$

(b) Find $\displaystyle\int x\cos 3x\,dx$

Hence, evaluate $\displaystyle\int_0^{\frac{\pi}{6}} x\cos 3x\,dx$　　　　　　　　[E]

69 Evaluate the integrals

(a) $\displaystyle\int_{\frac{1}{3}}^1 xe^{3x}\,dx$

(b) $\displaystyle\int_0^{\frac{\pi}{2}} \frac{\cos x}{1 + \sin x}\,dx$

(c) $\displaystyle\int_0^1 \frac{1-x}{(1+x)^2}\,dx$　　　　　　　　[E]

70 Express $\dfrac{1}{(3t+1)(t+3)}$ in partial fractions.

Use the substitution $t = \tan x$ to show that

$$\int_0^{\frac{\pi}{4}} \frac{1}{3 + 5\sin 2x}\,dx = \int_0^1 \frac{1}{(3t+1)(t+3)}\,dt$$

Hence show that $\displaystyle\int_0^{\frac{\pi}{4}} \frac{1}{3 + 5\sin 2x}\,dx = \tfrac{1}{8}\ln 3$　　　　　　　　[E]

71 (a) By using a suitable substitution, find

$$\int_1^2 x(x-1)^r\,dx,\, r > 0$$

(b) Sketch the curve C with equation $y^2 = (x-1)^3$, showing clearly the behaviour of the curve near the point with coordinates $(1, 0)$.

(c) Find the area of the finite region R bounded by C and the straight line $x = 2$.

72 The figure shows the curve with equation

$$y = xe^{-\frac{x}{2}}$$

(a) Determine the coordinates of the turning point.
The region R is bounded by the curve, the x-axis
and the line $x = 2$. Find
(b) the area of R
(c) the volume generated when R is rotated through
2π about the x-axis.   [E]

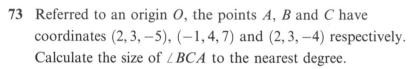

73 Referred to an origin O, the points A, B and C have
coordinates $(2, 3, -5)$, $(-1, 4, 7)$ and $(2, 3, -4)$ respectively.
Calculate the size of $\angle BCA$ to the nearest degree.

74 Referred to an origin O, the points P and Q are given by

$$\overrightarrow{OP} = (-2\mathbf{i} + 3\mathbf{j} + \mathbf{k}) \text{ and } \overrightarrow{OQ} = (3\mathbf{i} - 2\mathbf{j} + \mathbf{k})$$

(a) Show that $\overrightarrow{OP}.\overrightarrow{OQ} = -11$.
(b) Find the size of $\angle POQ$ to the nearest degree.

75 The curve C shown in the figure is given by

$$x = t + \frac{1}{t}, \quad y = t - \frac{1}{t}, \quad t > 0$$

where t is a parameter.

(a) Find $\dfrac{dy}{dx}$ in terms of t and deduce that the tangent to
C at the point $A(2, 0)$ has equation $x - 2 = 0$.
(b) For every point (x, y) on C, show that $x^2 - y^2 = 4$.
The point B has coordinates $(2\frac{1}{2}, 1\frac{1}{2})$ and the point D has
coordinates $(5, 0)$. The region R is bounded by the lines AD and
BD and arc AB of the curve C. The region R is rotated through
2π radians about the x-axis to form a solid of revolution.
(c) Find the volume of the solid, leaving your answer in
terms of π.   [E]

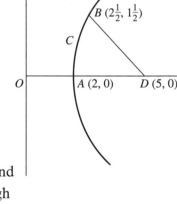

76 Express $\dfrac{1}{(1+x)(3+x)}$ in partial fractions.

Hence find the solution of the differential equation

$$\frac{dy}{dx} = \frac{y}{(1+x)(3+x)}, \quad x > -1$$

given that $y = 2$ at $x = 1$.
Express your answer in the form $y^2 = \mathrm{f}(x)$.   [E]

77 The line l_1 is parallel to the vector $(2\mathbf{i} - \mathbf{j})$ and passes through the point with position vector $(\mathbf{i} + \mathbf{j} - \mathbf{k})$. The line l_2 passes through the points with position vectors $(\mathbf{i} + \mathbf{j} - \mathbf{k})$ and $(3\mathbf{i} - 2\mathbf{j} + \mathbf{k})$. Find, to the nearest degree, the acute angle between l_1 and l_2.

78 A curve is defined in terms of the parameter t by the equations

$$x = t^3, \quad y = 3t^2$$

The points P and Q have parameters $t = 2$ and $t = 3$ respectively on this curve and O is the origin. Show that the gradient of the line OQ is equal to the gradient of the tangent to the curve at P.
Obtain an equation of the normal to the curve at Q.
Calculate the area of the finite region bounded by the curve and the line OQ.   [E]

79 Sketch the curve defined with parameter t by the equations

$$x = 1 + t, \quad y = 1 - t^2$$

Find the area of the finite region R enclosed by the curve and the x-axis. Find also the volume of the solid generated when R is rotated through 4 right angles about the x-axis. Show that the volume of the solid generated when R is rotated through 2 right angles about the line $x = 1$ is $\dfrac{\pi}{2}$.   [E]

80 Evaluate exactly:

(a) $\displaystyle\int_0^{\frac{\pi}{4}} \sin^2 x \cos^2 x \, dx$ (b) $\displaystyle\int_{-\frac{\pi}{4}}^{\frac{\pi}{4}} \cos 2x \cos 5x \, dx$

(c) $\displaystyle\int_0^{\frac{\pi}{2}} \cos^3 x \, dx$ (d) $\displaystyle\int_0^{\frac{\pi}{4}} \sin 3x \cos 2x \, dx$

Examination style paper

P3

1. The angle between the vectors $\mathbf{i} + \mathbf{j}$ and $2\mathbf{i} + \mathbf{j} + \lambda\mathbf{k}$ is $\dfrac{\pi}{4}$. Find the possible values of λ. **(5 marks)**

2. Find, in cartesian form, an equation of the circle which passes through the points $(2, 0)$, $(8, 0)$ and $(10, 4)$. **(4 marks)**

   Prove that the y-axis is a tangent to the circle and state the coordinates of the point of contact. **(2 marks)**

3. Solve the differential equation

$$\frac{\mathrm{d}y}{\mathrm{d}x} = 8y^3 \sin^2 x$$

   given that $y = \frac{1}{2}$ at $x = \dfrac{\pi}{4}$. **(7 marks)**

4. The quadratic function $\mathrm{f}(x)$ where $x \in \mathbb{R}$ has the following properties:
   (i) it has a remainder of 2 when divided by x,
   (ii) it has a remainder of 3 when divided by $x - 1$,
   (iii) it has a remainder of 10 when divided by $x - 2$.

   Find $\mathrm{f}(x)$ explicitly and hence deduce the range of f. **(9 marks)**

5. Referred to an origin O, the lines l_1 and l_2 have equations
   $l_1 \colon \mathbf{r} = \mathbf{i} + 2\mathbf{j} + 6\mathbf{k} + \lambda(\mathbf{i} + \mathbf{j} - 9\mathbf{k})$,
   $l_2 \colon \mathbf{r} = 4\mathbf{i} - 2\mathbf{j} - 8\mathbf{k} + \mu(\mathbf{i} - 6\mathbf{j} + 4\mathbf{k})$.
   (a) Prove that the lines intersect and find the position vector of the intersection point P. **(8 marks)**
   (b) Find the distance between O and P. **(2 marks)**

6. (i) Differentiate with respect to x:

   (a) $\cos^2 2x$,   (b) $\dfrac{\mathrm{e}^{-3x}}{1 + \tan x}$ **(5 marks)**

   (ii) A curve has equation $yx^2 - 3x = \ln y$. Find an equation of the normal to the curve at the point $(3, 1)$. **(6 marks)**

7. (i) Find $\int \tan^2 \frac{1}{2} x \, dx$. **(3 marks)**

(ii) Use the substitution $u = x^3 + 1$ to evaluate

$$\int_0^2 \frac{x^2}{\sqrt{(x^3 + 1)}} \, dx.$$ **(4 marks)**

(iii) Using integration by parts, or otherwise, find

$$\int x(3x + 1)^{-\frac{1}{2}} \, dx.$$

Hence evaluate $\int_1^5 x(3x + 1)^{-\frac{1}{2}} \, dx$. **(6 marks)**

8. $$f(x) \equiv \frac{1 - 4x - x^2}{(2 + x)(1 + x^2)}.$$

(a) Express $f(x)$ in partial fractions. **(4 marks)**

(b) Expand $f(x)$ in ascending powers of x up to and including the term in x^3, stating the set of values of x for which your series is valid. **(6 marks)**

(c) Prove that

$$\int_0^1 f(x) \, dx = \ln \frac{3}{4}$$ **(4 marks)**

Answers

Edexcel accepts no responsibility whatsoever for the accuracy or method of working in the answers given for examination questions.

Exercise 1A

1 $\dfrac{1}{x+2}+\dfrac{1}{x+3}$

2 $\dfrac{1}{x-1}+\dfrac{1}{x+3}$

3 $\dfrac{3}{x+4}-\dfrac{2}{x+3}$

4 $\dfrac{5}{x+2}-\dfrac{4}{x+3}$

5 $\dfrac{1}{x-1}+\dfrac{2}{2x-1}-\dfrac{1}{x+3}$

6 $\dfrac{x+1}{x^2+4}+\dfrac{2}{x-2}$

7 $\dfrac{1}{x-3}-\dfrac{2}{x^2+3}$

8 $\dfrac{2}{x^2+5}-\dfrac{2}{2x+3}$

9 $\dfrac{1}{5+2x^2}-\dfrac{3}{x+3}$

10 $\dfrac{2}{2x+1}-\dfrac{2x+1}{x^2+2x+7}$

11 $\dfrac{2}{x-5}+\dfrac{3}{(x-5)^2}$

12 $\dfrac{1}{x+3}-\dfrac{2}{(x+3)^2}+\dfrac{4}{(x+3)^3}$

13 $\dfrac{4}{(x-1)^2}-\dfrac{3}{x+2}$

14 $\dfrac{4}{(x-1)^2}-\dfrac{2}{x-1}-\dfrac{3}{x+2}$

15 $\dfrac{1}{2x+1}-\dfrac{1}{2x+3}+\dfrac{3}{(2x+3)^2}$

16 $1+\dfrac{1}{x-1}$

17 $x+1+\dfrac{1}{x-1}$

18 $1+\dfrac{1}{x-1}-\dfrac{1}{x+1}$

19 $1-\dfrac{2}{x}+\dfrac{3}{x-1}$

20 $x+\dfrac{1}{2(x-1)}+\dfrac{1}{2(x+1)}$

21 $2+\dfrac{3}{x+1}+\dfrac{1}{x-2}$

22 $2-\dfrac{1}{1-x}-\dfrac{3}{1+2x}$

23 $2x+\dfrac{1}{x+2}-\dfrac{1}{x+3}$

24 $x+1+\dfrac{3}{x-2}-\dfrac{1}{x+2}$

25 $-x+\dfrac{1}{x}-\dfrac{2}{x^2}+\dfrac{1}{x+1}$

26 $\dfrac{2}{2x-3}-\dfrac{3}{3x+2}$

27 $\dfrac{2}{2x-1}+\dfrac{1}{x+2}-\dfrac{3}{(x+2)^2}$

28 $\dfrac{1}{x^2+1}+\dfrac{1}{x+1}+\dfrac{2}{(x+1)^2}$

29 $2x+\dfrac{2}{2x+3}-\dfrac{1}{x^2}+\dfrac{1}{x}$

30 $\dfrac{2}{2x-1}-\dfrac{1}{3x+1}$

31 $\dfrac{3}{x-1}-\dfrac{1}{(x-1)^2}+\dfrac{3-2x}{x^2+2}$

32 $\dfrac{11}{7(x-2)}-\dfrac{1}{x+2}+\dfrac{13}{7(2x+3)}$

33 $1-\dfrac{1}{x}-\dfrac{x}{x^2+x+1}$

34 $\dfrac{3}{x^2}-\dfrac{1}{x}+\dfrac{2}{x+1}$

35 $\dfrac{1}{x+1}+\dfrac{-x+2}{x^2-x+1}$

Exercise 1B

1 $(x+2)(x+3)(x-2)$

2 $(x-1)(x-2)(x+4)$

3 $(2x-1)(x+3)(x-2)$

4 $(x-2)(2x+5)(3x-2)$

5 $(x-1)(3x-1)(x+7)$

6 21   **7** 448   **8** -154 **9** $-6\frac{3}{4}$

10 $-2\frac{1}{2}$ **11** 2   **12** 7   **13** 3   **14** 3

15 2   **16** $a = 1, b = -4$

17 $a = 2, b = 1$

18 $a = -4$ $(x + 1)(3x - 1)(x - 2)$

19 $a = 9, x = 1, -\frac{3}{2}, -4$

20 $a = 2, b = -11, x = -2, \frac{1}{2}, 3$

21 $a = 1, b = -6$ $(x^2 + 1)(x - 2)(x + 3)$

22 $p = -3, q = -11$ $(2x - 1)(x - 3)(x + 2)$

23 $k = -6, p(x) \equiv (x - 3)^2 + 2 > 0$
for all real x

24 $A = 5, B = -4; x = 1, -\frac{1}{2}, -3$

25 (a) $A = 8, B = -3$ (b) $\pm\frac{1}{2}, -\frac{2}{3}$

Exercise 1C

1 $1 - 2x + 3x^2 - 4x^3 + \ldots,$ $|x| < 1$

2 $1 + 3x + 6x^2 + 10x^3 + \ldots,$ $|x| < 1$

3 $1 + 5x + 15x^2 + 35x^3 + \ldots,$ $|x| < 1$

4 $1 - \frac{1}{2}x + \frac{3}{8}x^2 - \frac{5}{16}x^3 + \ldots,$ $|x| < 1$

5 $1 + \frac{3}{2}x + \frac{3}{8}x^2 - \frac{1}{16}x^3 \ldots,$ $|x| < 1$

6 $1 - \frac{3}{4}x - \frac{3}{32}x^2 - \frac{5}{128}x^3 \ldots$ $|x| < 1$

7 $1 - x - x^2 - \frac{5}{3}x^3 - \ldots$ $|x| < \frac{1}{3}$

8 $1 - x + 2x^2 - \frac{14}{3}x^3 + \ldots$ $|x| < \frac{1}{3}$

9 $1 + x + \frac{3}{4}x^2 + \frac{1}{2}x^3 + \ldots$ $|x| < 2$

10 $1 - 6x + 36x^2 - 216x^3 + \ldots$ $|x| < \frac{1}{6}$

11 $\frac{1}{3} - \frac{x}{9} + \frac{x^2}{27} - \frac{x^3}{81} + \ldots$ $|x| < 3$

12 $\frac{1}{4} + \frac{1}{4}x + \frac{3}{16}x^2 + \frac{1}{8}x^3 + \ldots$ $|x| < 2$

13 $2 + \frac{3}{4}x - \frac{9}{64}x^2 + \frac{27}{512}x^3 - \ldots$ $|x| < \frac{4}{3}$

14 $2 - \frac{5}{12}x - \frac{25}{288}x^2 - \frac{625}{20736}x^3 - \ldots$ $|x| < \frac{8}{5}$

15 $\frac{1}{10} - \frac{1}{2000}x + \frac{3}{800\,000}x^2 - \frac{1}{32\,000\,000}x^3 + \ldots$
                         $|x| < 100$

16 $2 + 3x + 5x^2 + 9x^3 + \ldots$ $|x| < \frac{1}{2}$

17 $3 - 3x + 9x^2 - 15x^3 + \ldots$ $|x| < \frac{1}{2}$

18 $-\frac{1}{4} - \frac{x}{16} - \frac{3}{64}x^2 - \frac{5}{256}x^3 - \ldots$ $|x| < 2$

19 $\frac{1}{2} - \frac{3}{4}x + \frac{7}{8}x^2 - \frac{15}{16}x^3 + \ldots$ $|x| < 1$

20 $-\frac{4}{3} + \frac{7}{18}x - \frac{31}{108}x^2 + \frac{73}{648}x^3 - \ldots$ $|x| < 2$

21 $1 + 3x + \frac{9}{2}x^2 + \frac{13}{2}x^3 + \ldots$

22 $1 - x^{-1} - \frac{1}{2}x^{-2} - \frac{1}{2}x^{-3}, \; 9.949\,874\,4;$
10.049\,875\,6

23 (a) $p = 5, q = -2$ (b) $-500, 3125$
     (c) $|x| < \frac{1}{5}$

24 $1 + \frac{2}{3}x + \frac{2}{9}x^2$ **25** $p = -2, q = -1$

26 (a) $p = -4, q = \frac{3}{2}$ (b) $4x^3, 6x^4$
     (c) $|x| < \frac{1}{4}$

Exercise 2A

1 $2(x + 2)$ **2** $7(x - 5)^6$

3 $-4(3 - x)^3$ **4** $6(2x - 1)^2$

5 $12(3x - 7)^3$ **6** $-2(x - 4)^{-3}$

7 $3(3 - x)^{-4}$ **8** $-(x - 7)^{-2}$

9 $3(5 - x)^{-2}$ **10** $48x(6x^2 + 1)^3$

11 $(x - 2)^{-1}$ **12** $2xe^{x^2}$

13 $-e^{3-x}$ **14** $-2x(6 - x^2)^{-1}$

15 $\dfrac{-(2x - 7)}{(x^2 - 7x + 2)^2}$

16 $\dfrac{-72x^2}{(3 - 4x^3)^2}$ **17** $\dfrac{2}{\sqrt{(5 + 4x)}}$

18 $-2x^{-\frac{1}{2}}(1 - x^{\frac{1}{2}})^3$ **19** $-2x^{\frac{1}{3}}(x^{\frac{2}{3}} + 5)^{-4}$

20 $(2x - 3)e^{x^2 - 3x}$ **21** $-4x^{-2}\left(1 + \dfrac{1}{x}\right)^3$

22 $-3(2x + x^{-2})\left(x^2 + \dfrac{1}{x}\right)^{-4}$

23 $-x(1 - x^2)^{-\frac{1}{2}}$ **24** $\dfrac{-4x}{(1 + x^2)^3}$

25 $\dfrac{6x - 4}{3x^2 - 4x + 3}$

26 $(\frac{1}{2}x^{-\frac{1}{2}} - \frac{1}{2}x^{-\frac{3}{2}})(x^{\frac{1}{2}} + x^{-\frac{1}{2}})^{-1}$ or $\dfrac{x - 1}{2x(x + 1)}$

27 $\dfrac{e^x}{e^x + 5}$ **28**   1   **29** $\dfrac{-54x^2}{(6x^3 - 5)^4}$

30 $\dfrac{-1}{x(\ln x)^2}$ **31** $-\frac{1}{108}$ **32** $\frac{7}{5}$

34 $\frac{28}{27}$ **35** $-\frac{4}{243}$

Exercise 2B

1 $x^2(5x - 3)(x - 1)$

2 $2x^3(7x + 2)(2x + 1)^2$

3 $xe^x(2 + x)$

4 $2 \ln x + \dfrac{2x - 1}{x}$

5 $2e^{2x} \ln x + \dfrac{e^{2x}}{x}$

6 $\ln(3x - 1) + \dfrac{3(x + 1)}{3x - 1}$

7 $(12x - 1)(2x - 1)^4$

8 $2(2x - 3)^2(x^2 + 1)(7x^2 - 6x + 3)$

9 $(1 + 9x^2)(x^2 + 1)^3$

10 $\dfrac{4x^2 - 3x - 2}{\sqrt{(x^2 - 1)}}$ **11** $\frac{5}{2}x^{\frac{3}{2}} - 3x^{\frac{1}{2}} + \frac{1}{2}x^{-\frac{1}{2}}$

12 $\frac{2}{3}(x - 2)^{-\frac{1}{3}}(x + 2)^{\frac{3}{4}} + \frac{3}{4}(x + 2)^{-\frac{1}{4}}(x - 2)^{\frac{2}{3}}$

13 $\dfrac{20x^2 - 12x - 2}{\sqrt{(4x - 3)}}$

14 $\frac{1}{2}x^{-\frac{1}{2}}\ln(x^2 - 1) + \dfrac{2x^{\frac{3}{2}}}{x^2 - 1}$

15 $(x - 1)^2 e^{x^2 - 2x}[2x^2 - 4x + 5]$

16 $\dfrac{-1}{(x - 1)^2}$ **17** $-1 + \dfrac{1}{(1 - x)^2}$

18 $\dfrac{6x^2 - x^4}{(2 - x^2)^2}$ **19** $\dfrac{e^x(2x - 1)}{(2x + 1)^2}$

20 $-3x^{-4}\ln(1 - x^2) - \dfrac{2x^{-2}}{1 - x^2}$

21 $\dfrac{6x^2}{(x^3 + 1)^2}$ **22** $\dfrac{x}{(x - 1)^{\frac{3}{2}}} - \dfrac{2}{}$

23 $\dfrac{2x - 2 - 2x^2}{(x^2 - 1)^2}$

24 $\dfrac{1}{2(x^4 + 1)\sqrt{x}} - \dfrac{4x^{\frac{7}{2}}}{(x^4 + 1)^2}$

25 $\dfrac{6e^x}{(3e^x + 1)^2}$

26 $\dfrac{2x(2x + 1)e^{x^2}\ln(2x + 1) - 2e^{x^2}}{(2x + 1)[\ln(2x + 1)]^2}$

27 $\dfrac{2x(x^2 - 1) - 2x(x^2 + 1)\ln(x^2 + 1)}{(x^2 + 1)(x^2 - 1)^2}$

28 $\dfrac{2x^5 + 4x^3 - 2x}{(x^2 + 1)^2}$

29 $\dfrac{6x^2(4x^3 - 1)^{-\frac{1}{2}}\ln(2x - 3) - 2(2x - 3)^{-1}(4x^3 - 1)^{\frac{1}{2}}}{[\ln(2x - 3)]^2}$

30 $\dfrac{2x(x^2 + 1)^{-1}(e^x + 1) - e^x\ln(x^2 + 1)}{(e^x + 1)^2}$

31 ± 1

32 $\dfrac{x}{(1 + x^2)^{\frac{1}{2}}}; \; 0, \frac{3}{5}$

33 $(0, 0), (3, 0), (\frac{9}{5}, 8.40)$

34 $\frac{1}{48}$ **35** $6 + \ln 4$

Exercise 2C

1 $2\,\text{cm}^2\,\text{s}^{-1}$

2 $0.0244\,\text{m}\,\text{s}^{-1}$

3 $112\,\text{cm}^2\,\text{s}^{-1}$

4 (a) $2.88\,\text{cm}^3\,\text{s}^{-1}$ (b) $2.88\,\text{cm}^2\,\text{s}^{-1}$

5 (a) $0.004\,77\,\text{cm}\,\text{s}^{-1}$ (b) $2.4\,\text{cm}^2\,\text{s}^{-1}$

6 7.78

7 (a) e

 (b) $A(1, 0)$, $B(e, e^{-1})$, $C(e^{\frac{3}{2}}, \frac{3}{2}e^{-\frac{3}{2}})$

9 stationary values at $x = -\frac{1}{2}, x = 1$;

 $\max y = -2, \min y = -8$

10 $0.005\,15\,\text{cm}\,\text{s}^{-1}$

11 $\dfrac{1}{x - 2} + \dfrac{1}{x + 2}, -\dfrac{5}{18}, 0$ **12** $-\dfrac{2}{125}$

13 $\dfrac{e}{4}, \dfrac{e}{4}$

17 $4y - 3x = 7$, $3y + 4x = 24$

18 (a) $A(-2, 0), B(0, \sqrt{2})$ (b) $2y - x = 3$

 (c) $\dfrac{3\sqrt{5}}{2}$ (d) $1 + \sqrt{2}$

20 $y = 2x - e$, $2y + x = 3e$, $\dfrac{5e}{2}, \dfrac{5e^2}{4}$

22 $3y + 4x = 10$, $3y + 10x = 95$

23 $(\frac{1}{2}, -\frac{1}{12})$ minimum, $(-\frac{1}{2}, \frac{1}{12})$ maximum

24 $1, -6, 9$

25 (a) tangent: $y = 4x - 1$, normal: $x + 4y = 4\frac{1}{2}$

 (b) $(9, -\frac{9}{8})$

26 (a) $(3, \frac{1}{12})$ (b) $(6, \frac{2}{27})$

27 (b) $(1, 1)$ maximum, $(-1, -1)$ minimum

28 (a) $(e^{-\frac{1}{2}}, -\frac{1}{2}e^{-1})$ (b) $y = 3ex - 2e^2$

29 $\dfrac{2}{x + 1} - \dfrac{1}{x^2 + 1}, 0$

30 $A = 2, B = 1, C = -1, 2, -3\frac{7}{8}$

Exercise 2D

1 $3\cos 3x$ **2** $\frac{1}{2}\cos\frac{1}{2}x$ **3** $-4\sin 4x$

4 $-\frac{2}{3}\sin\frac{2}{3}x$ **5** $2\sec^2 2x$ **6** $\frac{1}{4}\sec^2\dfrac{x}{4}$

7 $5\sec 5x\tan 5x$ **8** $-\frac{1}{2}\operatorname{cosec}\dfrac{x}{2}\cot\dfrac{x}{2}$

9 $-6\operatorname{cosec}^2 6x$ **10** $\frac{1}{2}\sec\dfrac{x}{2}\tan\dfrac{x}{2}$

11 $-\frac{3}{2}\operatorname{cosec}^2\dfrac{3x}{2}$

12 $-\frac{2}{3}\operatorname{cosec}\frac{2x}{3}\cot\frac{2x}{3}$

13 $2\sin x\cos x$ **14** $3\sin^2 x\cos x$

15 $\dfrac{\cos x}{2\sqrt{(\sin x)}}$ **16** $-4\cos^3 x\sin x$

17 $-5\cos^4 x\sin x$

18 $-\frac{1}{3}\sin x(\cos x)^{-\frac{2}{3}}$ **19** $2\tan x\sec^2 x$

20 $\dfrac{\sec^2 x}{2\sqrt{(\tan x)}}$ **21** $-\operatorname{cosec}^2 x$ **22** $\dfrac{-2\cos x}{\sin^3 x}$

23 $\dfrac{16\sin x}{\cos^5 x}$ **24** $-2\operatorname{cosec}^2 x\cot x$

25 $4\sin 2x\cos 2x$ **26** $-6\cos 3x\sin 3x$

27 $6\tan^2 2x\sec^2 2x$ **28** $4\sec^2 2x\tan 2x$

29 $3\cos(3x+5)$

30 $-4\cos(2x-4)\sin(2x-4)$

31 $-4\tan(1-2x)\sec^2(1-2x)$

32 $\dfrac{-6\cos 3x}{\sin^3 3x}$

33 $-8\operatorname{cosec}^2 4x\cot 4x$ **34** 0

35 $2(\cot x-\operatorname{cosec} x)(-\operatorname{cosec}^2 x+\operatorname{cosec} x\cot x)$

36 $-2\cos 2x$

37 $2\sin x\cos^2 x-\sin^3 x$

38 $\cos^4 x-3\sin^2 x\cos^2 x$

39 $\sec^3 x+\tan^2 x\sec x$

40 $\sec x\tan x\operatorname{cosec} x-\sec x\operatorname{cosec} x\cot x$

41 $\sin^2 x+2x\sin x\cos x$

42 $2x\cos x-x^2\sin x$

43 $2x\tan 2x+2x^2\sec^2 2x$

44 $\dfrac{\sin x-x\cos x}{\sin^2 x}$ **45** $\dfrac{x\cos x-\sin x}{x^2}$

46 $\dfrac{2x\cos x+x^2\sin x}{\cos^2 x}$

47 $\dfrac{-3(x\sin 3x+\cos 3x)}{x^4}$

48 $2\sec^2 2x\sec 3x+3\tan 2x\tan 3x\sec 3x$

49 $2\sin x\cos^4 x-3\cos^2 x\sin^3 x$

50 $e^x(\sin x+\cos x)$

51 $\dfrac{2e^{2x}(\cos x+\sin x)}{\cos^3 x}$

52 $2\sin x\cos x\ln x+\dfrac{\sin^2 x}{x}$

53 $\dfrac{e^{2x}(3\sin x+\cos x)}{(\sin x+\cos x)^2}$

54 $\dfrac{2\cos 3x+6(2x-5)\sin 3x\ln(2x-5)}{(2x-5)\cos^3 3x}$

57 $\frac{3}{7}$

58 $y-0.6=0.8(x-0.644)$ is tangent;
$y-0.6=-1.25(x-0.644)$ is normal

59 2 **60** $\dfrac{-4\sin 2x}{(1-\cos 2x)^2}$

63 $\dfrac{\pi}{2}-1$ **64** 1.00

65 $y-1=-\frac{1}{2}x$

66 $\frac{2\sqrt{3}}{27}\pi a^3$

67 maximum at $(1.11, 4.09)$;
minimum at $(4.25, -2190)$

Exercise 2E

1 $\dfrac{1}{y}$ **2** $-\dfrac{x}{y}$ **3** $\dfrac{-y}{2x}$

4 $\dfrac{-(x+y)}{x+3y}$ **5** $\cot x\cot y$

6 $-\dfrac{y}{x}$ **7** -1 **8** $\dfrac{-y\ln y}{x\ln x}$

9 $y(e^{-x}-\ln y)$ **10** $\dfrac{-y}{x+2y}$

11 $\dfrac{1}{3t}$ **12** $\dfrac{3t}{2}$ **13** $-\dfrac{2}{t^3}$

14 $-\dfrac{b}{a}\cot t$ **15** $\operatorname{cosec} t$

16 $\dfrac{\sin t+\cos t}{\cos t-\sin t}$ **17** $\dfrac{\sin t+t\cos t}{\cos t-t\sin t}$

18 $\dfrac{\cos\theta}{1+\sin\theta}$ **19** $1+e^{-2u}$ **20** $\cot 2\theta$

21 $y-4=4\ln 2(x-2),\ y-32=32\ln 2\,(x-5)$;
$\dfrac{38\ln 2-7}{7\ln 2}$

22 $4y-9x=34$

23 tangent: $2y=3-5x$;
normal: $5y=2x-7,\ x^5=y^2$

24 $3y=12x+4$ **25** $(2, 3)$

26 $y=2x-3$

27 $2t-t^2, 6y+8x=11$ **28** $-\frac{40}{33}$

29 (a) $10^x\ln 10$ (b) $2x(2^{x^2})\ln 2$
     (c) $-5^{-x}\ln 5$

30 $2y+2x-\pi=0$

Exercise 3A

1. (a) $(2, 3), 5$ (b) $(-3, -2), 7$
   (c) $(\frac{1}{2}, \frac{1}{2}), 1$ (d) $(5, -7), 4$
   (e) $(\frac{1}{5}, -\frac{6}{5}), \frac{2}{5}$

2. (a) $x^2 + y^2 = 36$ (b) $x^2 + (y + 7)^2 = 4$
   (c) $(x - 3)^2 + (y + 4)^2 = 25$
   (d) $(x + \frac{1}{2})^2 + (y + \frac{1}{3})^2 = 1$
   (e) $(x + 2)^2 + (y + 3)^2 = 10$

3. (a) $(x - 1)(x - 3) + y^2 = 0$
   (b) $x(x - 8) + y(y + 6) = 0$
   (c) $(x - 5)(x + 3) + (y - 5)(y + 3) = 0$

5. (a) $(x - 3)^2 + (y - 2)^2 = 13$
   (b) $(x - 2)^2 + (y - 1)^2 = 25$
   (c) $(x + 3)^2 + (y - 2)^2 = 25$

6. (a) $(x - 3)(x - 2) + (y - 2)(y - 1) = 0$
   (b) $x^2 + y^2 + 3x - 11y = 0$

7. (a) $3x - 4y + 25 = 0$
   (b) $2x - 7y + 3 = 0$
   (c) $4x - y + 2 = 0$

8. (a) $4x + 3y = 0$
   (b) $7x + 2y = 16$
   (c) $x + 4y = 8$

9. $8, 18$

10. $3x + 2y = 0$; $(8, -12)$

11. $\dfrac{9 \pm \sqrt{17}}{8}$

12. $x^2 + y^2 - 10x + 2y + 1 = 0$,
    $x^2 + y^2 - 2x - 6y + 9 = 0$; $4\sqrt{2}$

13. $3x^2 + 3y^2 + 6x + 8y - 25 = 0$

14. $y = 4, 4x + 3y = 20$

15. (a) $1, 21$ (b) $(0, 1)$, $(-10, 11)$

16. $(3\frac{1}{2}, -2), 6\frac{1}{2}$

17. $(x - 1)(x - 4\frac{1}{2}) + (y - 2)(y - 4) = 0$

18. $(6, 8), (-1, 7), (0, 0)$; $x^2 + y^2 - 6x - 8y = 0$

Exercise 3C

21. a; $(a, 3a)$ 22. 3; $(2, -4)$

24. (a) $\left[\dfrac{2a + a\cos\theta}{2}, \dfrac{4a + a\sin\theta}{2}\right]$,
    (b) $(x - a)^2 + (y - 2a)^2 = \frac{1}{4}a^2$

Review exercise 1

1. $\dfrac{dy}{dx} = \dfrac{2e^{2x} - y}{x - 2e^{2y}}$

2. (a) $4a + 2b + 14$ (b) $9a - 3b - 21$
   $a = b + 7$

3. $1 - x - \frac{3}{2}x^2 - \frac{7}{2}x^3 \ldots$, $|x| < \frac{1}{4}$

4. Centre $(8, 6)$ radius 2.
   Least distance is 8 and greatest distance is 12.

5. (a) -1; $(x + 1)(x - 2)(2x + 1)$ (b) 18

6. (a) 0.1 (b) 1820

7. (a) $e^{-x}(\cos x - \sin x)$ (b) $2\tan x$

8. $2.88 \, \text{cm}^2 \, \text{s}^{-1}$

9. $A = -4, B = 4$; $x = 1, \pm 2$

10. $(1, 1)$, $(3\frac{1}{3}, -3\frac{1}{3})$

11. $x^2 + y^2 - 6x - 16 = 0$

12. $1.5 \, \text{cm}^2 \, \text{s}^{-1}$

13. $A = B = \frac{2}{9}, C = \frac{2}{3}$

14. $\dfrac{1}{x - 2} - \dfrac{1}{x + 2}$; $-1 - \dfrac{x^2}{4}$

15. (a) (i) $\dfrac{2}{x}$ (ii) $2x \sin 3x + 3x^2 \cos 3x$
    (b) $\frac{1}{7}$

16. $e^{2x}(2\cos x - \sin x)$ (b) $y - 2x + 1$

17. $A = \frac{1}{3}, B = -\frac{1}{3}, C = \frac{2}{3}$

18. $(4, 3), 5$; $(12, 9), 5$

19. (a) 2 (b) $-\frac{4}{5}$

20. $\frac{2\sqrt{3}}{27} \pi l^3$

21. $\dfrac{1}{1 - 3x} + \dfrac{2x}{1 + 6x^2}$, $1 + 5x + 9x^2 + 15x^3$, $|x| < \frac{1}{3}$

22. (a) $\dfrac{2\cos 2x \cos 3x + 3\sin 2x \sin 3x}{\cos^2 3x}$
    (b) $5\tan^4 x \sec^2 x$
    (c) $-\dfrac{2x}{3y^2}$; $y - 2 = -\frac{1}{3}(x - 2)$

23. (a) (i) $\dfrac{-(x^3 + 4)}{2x^3(1 + x^3)^{\frac{1}{2}}}$
    (ii) $\dfrac{2\cos x - 3\sin x + 1}{(2 + \cos x)(3 - \sin x)}$

24. $\frac{4}{5} p \, \text{cm}^2 \, \text{s}^{-1}$

25. (a) $x^2 + y^2 = 9$ (b) $x^2 + 4y^2 = 16$
    (c) $x^2 - y^2 = 9$

26 $\dfrac{3}{1+x} + \dfrac{4-3x}{1+x^2}$; $7 - 6x - x^2 + 7x^4$

27 $y = x$

30 $\dfrac{2}{x+3} - \dfrac{1}{2x-1}$; $\dfrac{5}{3} + \dfrac{16x}{9} + \dfrac{110x^2}{27}$, $|x| < \frac{1}{2}$

31 $(0,5)$, radius $= 5$

33 $x^2 + y^2 + 4x + 10y - 71 = 0$

34 (ii) 3

35 (a) (i) 12   (ii) $(x+3)(5x-7)(x-4)$
    (b) 4

36 $1 + \dfrac{2}{x+1} + \dfrac{3}{x+2}$; $-\frac{5}{6}$

37 (b) $x + 3y + 5 = 0$ **38** $32\,°C$

39 (b) $(2x-1)(x+1)(x+5)$
    (c) $\pm1, -5$

40 (a) (i) $4, -5$ (ii) $2, 11$
    (b) (ii) $(x+1)(2x+5)(x-4)$
       (iii) $-2\frac{1}{2}, -1, 4$

41 $(8,2)$ **42** $p = -28$, $q = -16$

43 1.24×10^{-4}

44 $(-1,0)$; $1, -3$; $(\frac{1}{2})^{\frac{1}{3}}$, minimum

45 $a = 3$, $b = 1$, remainder 15

46 $A = 6$, $B = -2$, $C = 5$

48 (a) $\dfrac{x+1}{x-2}$, $x \in \mathbb{R}$, $x \neq 2$

49 min. $\frac{1}{2}$ at $x = \dfrac{\pi}{2}$, max. $\dfrac{11}{4}$ at $x = \dfrac{7\pi}{6}$ and $x = \dfrac{11\pi}{6}$

50 (a) -38, $(x+4)(3x-2)(x-3)$
    (b) -5

52 $9x - 8y + 18 = 0$

53 $c = 6$, $x = -2, \frac{1}{2}, 3$

54 (a) $(0, -\frac{1}{4})$ (b) $x = \pm2$, $y = 1$

56 $1 - 4x - 8x^2 - 32x^3 \ldots$; $|x| < \frac{1}{8}$

57 (a) False, $n = 4$, $f(4) = 21$, not prime.
    (b) True, $n(n+1)$ even, so $n^2 + n + 1$ odd.

58 (a) $A = 2$, $B = -1$, $C = 0$
    (b) $\dfrac{-4}{(1+2x)^2} + \dfrac{x^2 - 1}{(1+x^2)^2}$

59 (a) $\dfrac{x\cos x - \sin x}{x^2}$ (b) $\dfrac{-2x}{x^2 + 9}$

60 $1 + 4x - 6x^2 - 32x^3$; $a = 6$, $b = 4$; $|x| < \frac{1}{4}$

61 (b) 2

62 (a) $0, \dfrac{\pi}{2}, \pi$

    (b) $\left(\dfrac{3\pi}{8}, \dfrac{1}{\sqrt{2}} e^{\frac{3\pi}{4}}\right)$, $\left(\dfrac{7\pi}{8}, \dfrac{1}{\sqrt{2}} e^{\frac{7\pi}{4}}\right)$

    (d) max. at $x = \dfrac{3\pi}{8}$, min. at $x = \dfrac{7\pi}{8}$

63 (a) $x^2 + y^2 - 2x - 8y + 8 = 0$
    (b) Point is *inside* the circle.

64 (a) $p = \frac{1}{12}$, $q = -\frac{1}{288}$ (b) 2.41
    (c) 2.15%

66 (a) $3y - 2x = 1$ (b) $(8,4)$

67 (a) 13
    (b) $\dfrac{1}{105}\left[\dfrac{5}{x+2} + \dfrac{27}{3x-1} - \dfrac{28}{2x+1}\right]$
    (c) -6.75

68 $A = 2$, $B = -3$; $\frac{11}{2} - \frac{11}{4}x + \frac{13}{8}x^2$

69 (a) $\dfrac{dx}{dt} = -2\sin t + 2\cos 2t$,

        $\dfrac{dy}{dt} = -\sin t - 4\cos 2t$

    (b) $\frac{1}{2}$ (c) $y + 2x = \dfrac{5\sqrt{2}}{2}$

70 (a) $(3,-4), (-2,8)$ (b) $4, 9$
    (d) $x^2 + y^2 - x - 4y - 38 = 0$

71 (a) $xe^{-3x}(2-3x)$ (b) $\dfrac{4x}{1-x^4}$

72 $7x + \frac{1}{2}x^2 + \frac{43}{3}x^3 + \ldots$

73 (a) $\dfrac{3}{(x+1)^2} - \dfrac{2}{x+1} + \dfrac{2}{x+2}$ (b) $-\frac{17}{36}$

75 $t^3 y + 2x - 3t = 0$

76 (a) $1 - x - x^2 - \frac{5}{3}x^3$
    (b) $9.989\,989\,98$ (9 s.f.)

77 $x^2 + y^2 - x - 3y - 4 = 0$; $(0,-1)$, $(1,4)$

78 $1 + \dfrac{x}{5} - \dfrac{2x^2}{25} + \dfrac{6x^3}{125}$; $2.012\,35$

79 (a) $20\pi k\,\mathrm{m^2\,s^{-1}}$ (b) $25\pi k\,\mathrm{m^3\,s^{-1}}$

80 $-\frac{23}{6}$

Exercise 4A

[The constant of integration is omitted in indefinite integration.]

1 $\frac{1}{4}\sin 4x$ **2** $-\frac{1}{3}\cos 3x$ **3** $-2\cos\frac{1}{2}x$
4 $\frac{2}{3}\sin\frac{3}{2}x$ **5** $\tan x$ **6** $-\frac{1}{3}\cot 3x$

7 $-\frac{1}{2}e^{-2x}$ **8** $\frac{1}{3}e^{3x-2}$

9 $\frac{1}{2}\ln|2x-5|$

10 $\frac{1}{16}(4x-3)^4$ **11** $\frac{1}{5}\sin(5x+4)$

12 $\frac{1}{4}\cos(3-4x)$ **13** $-\frac{1}{6}(3-2x)^3$

14 $\dfrac{1}{2(3-2x)}$

15 $\frac{1}{2}e^{2x}-2x-\frac{1}{2}e^{-2x}$ **16** $4\tan\dfrac{x}{2}$

17 $\frac{1}{3}\sec 3x$ **18** $-\frac{1}{2}\operatorname{cosec} 2x$

19 $\tan x-\frac{1}{3}x^3$ **20** $x^2-\frac{1}{2}\cos 2x$

21 $\frac{1}{2}$ **22** $1\frac{1}{2}$ **23** 0 **24** 1

25 $1+\frac{1}{8}\pi^2$ **26** $2-2\sqrt{3}$ **27** $\frac{1}{3}\ln\frac{5}{2}$

28 10 **29** $\frac{1}{2}(e-e^{-3})$ **30** 2

Exercise 4B

[The constant of integration is omitted in indefinite integration.]

1 $\frac{1}{2}x-\frac{1}{4}\sin 2x$ **2** $-\cot x-x$

3 $\frac{1}{2}\tan 2x-x$

4 $\dfrac{3x}{2}+2\sin x+\frac{1}{4}\sin 2x$

5 $3x+4\cos x-\sin 2x$

6 $\dfrac{5x}{2}+\frac{1}{4}\sin 2x+\tan x$ **7** $\frac{1}{4}\ln\left|\dfrac{x-2}{x+2}\right|$

8 $\ln|x-3|-2\ln|x-2|$

9 $\ln|x-1|-2\ln|2x+1|$

10 $\frac{1}{2}\ln|2x+1|-\frac{1}{3}\ln|3x+1|$

11 $\ln\left|\dfrac{2x+1}{3x+1}\right|$ **12** $\ln\left|\dfrac{3+2x}{3-2x}\right|$

13 $A=1, B=\frac{1}{2}, C=-\frac{1}{2}$;

$x+\frac{1}{2}\ln\left|\dfrac{x-1}{x+1}\right|$

14 $\dfrac{\pi}{3}+\dfrac{1}{4}$ **15** 0

16 $\dfrac{\sqrt{3}}{4}+\dfrac{1}{3}$ **17** (a) $\dfrac{\pi}{2}-1$ (b) $\dfrac{\pi}{2}+1$

Exercise 4C

[The constant of integration is omitted in indefinite integration]

1 $\frac{1}{4}\sin^4 x$ **2** $\frac{1}{3}\tan^3 x$ **3** $\frac{1}{8}(x^2+1)^4$

4 $\frac{1}{6}(x^4-1)^{\frac{3}{2}}$ **5** $2\sqrt{(e^x-1)}$ **6** $\frac{1}{2}\sec^2 x$

7 $\frac{1}{2}e^{x^2}$ **8** $\frac{1}{3}(\ln|x|)^3$

9 $x+1-2\ln|x+1|-\dfrac{1}{x+1}$

10 $\frac{2}{3}(x-2)(x+1)^{\frac{1}{2}}$

11 $6\frac{2}{3}$ **13** $-\frac{1}{12}$

14 (a) $e-1$ (b) $\frac{1}{4}\sqrt{3}$ (c) $\ln\frac{5}{4}$

15 (a) $\frac{1}{2}\ln 2$ (b) $\frac{1}{2}\ln 2$

Exercise 4D

[The constant of integration is omitted in indefinite integration.]

1 $-e^{-x}(x+1)$ **2** $\frac{1}{3}xe^{3x}-\frac{1}{9}e^{3x}$

3 $-x\cos x+\sin x$ **4** $\dfrac{x^2}{2}\ln|x|-\dfrac{x^2}{4}$

5 $x\ln|x-1|-x-\ln|x-1|$

6 $\frac{1}{3}x\sin 3x+\frac{1}{9}\cos 3x$

7 $\dfrac{(5x+1)(x-1)^5}{30}$ **8** $\frac{2}{15}(3x+2)(x-1)^{\frac{3}{2}}$

9 $e^x(x^2-2x+2)$

10 $x^2\sin x+2x\cos x-2\sin x$

11 $-e^{-x}(x^2+2x+2)$

12 $\dfrac{x^4}{16}(4\ln x-1)$ **13** π

14 $\dfrac{\pi}{\sqrt{2}}+\dfrac{4}{\sqrt{2}}-4$ **15** $\frac{2}{9}e^3+\frac{1}{9}$

16 $-\frac{1}{20}$ **17** 8.4 **18** $\frac{1}{9}(1-4e^{-3})$

19 $e-2$ **20** $\frac{1}{2}(e^{\frac{\pi}{2}}+1)$

Exercise 4E

[The constant of integration is omitted in indefinite integration.]

1 $\frac{1}{6}(4x-5)^{\frac{3}{2}}$ **2** $\frac{1}{4}\ln|4x+5|$

3 $x-2\ln|x|-\dfrac{1}{x}$ **4** $\frac{1}{2}\sin^2 x$

5 $\frac{1}{3}\ln|\sec 3x|$ **6** $-\frac{1}{3}x\cos 3x+\frac{1}{9}\sin 3x$

7 $2x^{\frac{1}{2}}+\frac{2}{3}x^{\frac{3}{2}}$ **8** $x-\ln|x+1|$

9 $-\frac{1}{5}\cos^5 x$ **10** $3x\ln x-3x$

11 $3\ln|x-1|-2\ln|x|$

12 $-\frac{1}{2}(1+\tan x)^{-2}$ **13** $\frac{1}{2}x-\frac{1}{8}\sin 4x$

14 $\frac{1}{2}x^2+2x+4\ln|x-2|$

15 $\frac{5}{2}x+\frac{3}{4}\sin 2x-\cos 2x$

16 $2e^{\frac{1}{2}x}(x^2 - 4x + 8)$

17 $\frac{1}{4}\ln\left|\dfrac{x-2}{x+2}\right|$

18 $\frac{1}{18}\ln|9x^2 + 1|$

19 $x + 2x^{-1} - \frac{1}{3}x^{-3}$

20 $\frac{1}{3}(2 - 3x)^{-1}$

21 $-\frac{1}{5}\ln|4 - 5x|$

22 $\frac{1}{3}\ln|\sin 3x|$

23 $-\frac{1}{2}\operatorname{cosec} 2x$

24 $-\frac{1}{3}\cot 3x - x$

25 $\frac{1}{5}x\sin 5x + \frac{1}{25}\cos 5x$

26 $\frac{2}{3}(x+2)\sqrt{(x-1)}$

27 $-e^{-x}(x^2 + 2x + 2)$

28 $-\frac{2}{3}\cos^3 x + \cos x$

29 $-\frac{2}{3}\cos^3 x$

30 $\frac{1}{2}\sec 2x$

31 $x + \ln|1 + x^2|$

32 $\ln\left|\dfrac{x-4}{x-2}\right|$

33 $\dfrac{1}{x} + \ln\left|\dfrac{x-1}{x}\right|$

34 $x - \frac{1}{2}\cot 2x$

35 $x + 8\ln|x - 4|$

36 $\frac{1}{2}\ln\left|\dfrac{x^2 - 1}{x^2}\right|$

37 $\frac{1}{3}\ln|x^3 + 1|$

38 $\frac{1}{2}e^{2x} + \frac{1}{3}x^3 + 2xe^x - 2e^x$

39 $\frac{1}{16}x^4(4\ln x - 1)$

40 $\frac{1}{2}e^{x^2}(x^2 - 1)$

41 $\frac{1}{3}\cos^3 x - \cos x$

42 $\sin x - \frac{1}{3}\sin^3 x, \ -\cos x + \frac{2}{3}\cos^3 x - \frac{1}{5}\cos^5 x$

43 $\frac{1}{3}\tan^3 x - \tan x + x$

44 $\frac{1}{3}\tan^3 x + \tan x, \ x + \cot x - \frac{1}{3}\cot^3 x$

45 (a) $-\frac{1}{10}\cos 10x - \frac{1}{2}\cos 2x$

  (b) $-\frac{1}{3}\cos\dfrac{3x}{2} - \cos\dfrac{x}{2}$

46 $\frac{5}{2}\ln 3 - \ln 2$

48 $\frac{1}{3}\left[8 - \dfrac{8}{3\sqrt{3}}\right]$

49 $\ln 2 + \frac{1}{2}\ln 13 - \frac{1}{2}\ln 20$

50 $\frac{1}{2}\ln\frac{5}{2}$

Exercise 4F

1 $\dfrac{\sqrt{3} - 1}{2}$

2 1

3 $2e^3$

4 $5\ln 5 - 2\ln 2 - 3$

5 $\dfrac{\pi}{6}$

6 $2 - \dfrac{\pi}{4}$

7 0.02

8 $\frac{1}{2}(e^4 - e)$

9 $\ln\frac{4}{3}$

10 $\frac{26}{27}\sqrt{3}$

11 8π

12 168π

13 $\frac{1}{2}\pi^2$

14 $\pi\ln\frac{5}{2}$

15 $\dfrac{\pi}{4}e^2(3e^2 - 1)$

16 $347\frac{1}{15}\pi$

17 $\dfrac{\pi}{4}(4 - \pi)$

18 $\pi[3(\ln 3)^2 - 6\ln 3 + 4]$

19 $\dfrac{\pi}{4}[15 + 16\ln 2]$

20 $\frac{64}{15}\pi$

24 $60, \dfrac{16\,256\pi}{7}$

25 (a) $\sqrt{2} - 1$ (b) $\dfrac{\pi}{4}(\pi - 2)$

26 (a) $12\frac{2}{3}$ (b) $32\frac{1}{2}\pi$

27 (a) 27 (b) $\dfrac{1773\pi}{5}$

28 $16\ln\frac{16}{3}, \frac{208}{3}\pi$ 29 (a) $\dfrac{3\pi}{2}$ (b) 12π

Exercise 4G

1 $y = \frac{1}{2}e^{2x-1} + C$

2 $2x + e^{1-2y} = C$

3 $4y = 2x + \sin 2x + C$

4 $\tan y = x + C$

5 $2\ln y = x^2 + C$

6 $e^{-y} + e^x = C$

7 $\sin y = x\ln x - x + C$

8 $y + 2 = C(x + 1)$

9 $\ln y = \ln x + \dfrac{x^2}{2} + C$

10 $y^2 = 2\operatorname{cosec} x + C$

11 $6y = 2x^3 + 3x^2$

12 $3y = \sin^3 x - 1$

13 $\ln(3y + 1) = 3x - 3$

14 $2e^{-y} + x^2 - 3 = 0$

15 $2\sin y = \sec x$

16 $2\tan y = x - \sin x\cos x$

17 $\sin^2 y = \dfrac{11}{4} - \dfrac{2}{x}$

18 $y^2 = 2\tan x(\tan x + 1) + 5$

19 $y\cos y - \sin y = \dfrac{\pi}{2} - x\sin x - \cos x$

20 $\dfrac{y-1}{y+1} = \dfrac{2\sin^2 x}{3}$

22 (a) 19 minutes   (b) 53.08

23 26.6 days   24 $\ln y = \ln x + 1 - x^2$

25 $y = x\ln x - x + e + 1$

Exercise 5A

1 (a) $\overrightarrow{AF}$ (b) $\overrightarrow{PU}$ (c) $\overrightarrow{LP} + \overrightarrow{QR}$
   (d) $\overrightarrow{AH}$ (e) $\overrightarrow{PS}$

2 (a) $\mathbf{r} - \mathbf{q}$ (b) $\mathbf{q} - \mathbf{p}$ (c) $\mathbf{p} + \mathbf{r} - \mathbf{q}$

3 (a) $\mathbf{a} - \mathbf{d}$ (b) $-(\mathbf{b} + \mathbf{c} + \mathbf{d})$
   (c) $-(\mathbf{a} + \mathbf{b})$ (d) $\mathbf{b} + \mathbf{d}$

Exercise 5B

1. (a) 3**a**   (b) 2**a**
   (c) 3**b**   (d) −2**b**
   (e) −4**b**   (f) 2**a** + **b**
   (g) 3**a** + 2**b**   (h) **a** − **b**
   (i) −(**a** + 2**b**)   (j) 4(**a** + **b**)
   (k) 2(**b** − **a**)   (l) 3(**b** − **a**)
   (m) −5**a** + 4**b**

2. (a), (b), (e)

3. $\overrightarrow{AE} = 3\mathbf{b}$, $\overrightarrow{AC} = \mathbf{a} + \mathbf{b}$, $\overrightarrow{EC} = \mathbf{a} - 2\mathbf{b}$

4. (a) **a** − **b**   (b) $\frac{4}{5}(\mathbf{b} - \mathbf{a})$
   (c) $\frac{1}{5}\mathbf{a} + \frac{4}{5}\mathbf{b}$

5. (a) 2**b** − **a**   (b) $\mathbf{a} + \frac{1}{2}\mathbf{b}$
   (c) **a** − 3**b**   (d) $\mathbf{a} - 2\frac{1}{2}\mathbf{b}$

7. $\overrightarrow{OD} = \frac{1}{2}(\mathbf{b} + \mathbf{c})$, $\overrightarrow{OG} = \frac{1}{3}(\mathbf{a} + \mathbf{b} + \mathbf{c})$

8. (a) **a** + **b**   (b) **b** − **a**
   (c) **b** − 2**a**   (d) −(**a** + **b**)

10. (a) $\overrightarrow{AB} = 6(\mathbf{b} - \mathbf{a})$, $\overrightarrow{OP} = 2(\mathbf{a} + 2\mathbf{b})$,
    $\overrightarrow{MN} = 2(\mathbf{b} - \mathbf{a})$
    (c) 2.4

11. (a) $\lambda = 7, \mu = 1$ (b) $\lambda = -8, \mu = 2$
    (c) $\lambda = 6, \mu = 5$ (d) $\lambda = 2\frac{2}{3}, \mu = -5$
    (e) $\lambda = \frac{1}{2}, \mu = \frac{1}{2}$ (f) $\lambda = 2, \mu = -3$
    (g) $\lambda = -3, \mu = -9$ (h) $\lambda = 3, \mu = -1$

12. (a) $\overrightarrow{AB} = \mathbf{b} - \mathbf{a}$, $\overrightarrow{AC} = \frac{5}{6}(\mathbf{b} - \mathbf{a})$,
    $\overrightarrow{OC} = \frac{1}{6}\mathbf{a} + \frac{5}{6}\mathbf{b}$
    (b) $\overrightarrow{CE} = -\frac{1}{6}\mathbf{a} + (\lambda - \frac{5}{6})\mathbf{b}$
    (c) $\overrightarrow{ED} = -\mu\mathbf{a} + (\mu - \lambda)\mathbf{b}$
    (d) $\lambda = \frac{1}{2}, \mu = \frac{1}{6}$

13. (a) $\frac{2}{3}\mathbf{b}$ (b) $\frac{5}{6}\mathbf{b}$ (c) $\frac{1}{2}\mathbf{a} + \frac{1}{3}\mathbf{b}$
    (d) $(1 - t)\mathbf{a} + \frac{5}{6}t\mathbf{b}$ (e) $\frac{1}{2}s\mathbf{a} + (1 - \frac{2}{3}s)\mathbf{b}$
    (f) $\frac{1}{3}\mathbf{a} + \frac{5}{9}\mathbf{b}$ (g) $CP : CQ = 2 : 3$

14. (a) (i) 2**a** − **b** (ii) **a** − 3**b** (iii) 2**a** − 3**b**
    (b) $2\lambda\mathbf{a} + (1 - \lambda)\mathbf{b}$ (f) $\frac{4}{3}\mathbf{a} + \mathbf{b}$

15. (a) (i) (**b** − **a**) (ii) $\frac{2}{3}\mathbf{a} - \frac{7}{4}\mathbf{b}$
    (b) $\mu(\mathbf{b} - \mathbf{a}) + \frac{3}{4}\mathbf{b} = \lambda(-\frac{2}{3}\mathbf{a} + \frac{7}{4}\mathbf{b})$
    (d) $\frac{6}{13}$ (e) $\frac{6}{13}\mathbf{a} + \frac{7}{13}\mathbf{b}$ (f) $\frac{13}{10}$

16. (a) **b** − **a**   (b) $\frac{1}{2}(\mathbf{a} + \mathbf{b})$
    (c) $\frac{3}{8}(\mathbf{a} + \mathbf{b})$ (d) $-\frac{5}{8}\mathbf{a} + \frac{3}{8}\mathbf{b}$
    (e) −**a** + k**b**
    (f) $5 : 3; k = \frac{3}{5}$

17. (a) $\overrightarrow{OM} = \frac{1}{2}\mathbf{a}$, $\overrightarrow{OX} = \frac{3}{4}\mathbf{b}$, $\overrightarrow{OY} = \frac{1}{5}\mathbf{a} + \frac{3}{5}\mathbf{b}$
    (d) 2 : 3

18. (a) $\frac{1}{3}\mathbf{a}$ (b) $\frac{1}{4}\mathbf{a} + \frac{3}{4}\mathbf{b}$
    (c) $-\frac{1}{12}\mathbf{a} + \frac{3}{4}\mathbf{b}$ (d) 2**b**
    (e) $\frac{1}{4}(\mathbf{b} - \mathbf{a})$ (f) $2\frac{1}{4}\mathbf{b} - \frac{1}{4}\mathbf{a}$
    (g) 1 : 3
    (h) $\overrightarrow{EB} = -\frac{1}{3}\mathbf{a} + \mathbf{b}$, $\overrightarrow{AG} = 3\mathbf{b} - \mathbf{a}$

19. (a) $\overrightarrow{OB} = 5\mathbf{a} + 3\mathbf{b}$, $\overrightarrow{OC} = \frac{3}{2}(5\mathbf{a} + 3\mathbf{b})$
    $\overrightarrow{DC} = 4\frac{1}{2}\mathbf{a} + 4\frac{1}{2}\mathbf{b}$
    (b) $\mu\mathbf{b} = -2\mathbf{a} + \lambda(\mathbf{a} + \mathbf{b})$
    (c) $\lambda = \mu = 2$
    (d) (i) 2 : 1 (ii) 4 : 5
    (e) $\frac{10}{3}\mathbf{a} = 2\mathbf{b}$

20. (a) (i) $\frac{2}{3}\mathbf{a}$ (ii) $\frac{1}{3}\mathbf{b}$ (iii) $\frac{1}{3}\mathbf{b} - \frac{2}{3}\mathbf{a}$
    (b) **b** − 2**a**   (c) 3 : 1
    (d) $\overrightarrow{PS} = \frac{2}{3}\mathbf{a} - \frac{1}{3}\mathbf{b}$, $\overrightarrow{AS} = \frac{1}{3}(\mathbf{a} - \mathbf{b})$

Exercise 5C

1. (a) 13   (b) 25   (c) $\sqrt{26}$ (d) 7
   (e) 3   (f) $\sqrt{54}$

2. (a) $\frac{1}{13}(-5\mathbf{i} + 12\mathbf{j})$ (b) $\frac{1}{\sqrt{26}}(-\mathbf{i} - 5\mathbf{j})$
   (c) $\frac{1}{\sqrt{65}}(4\mathbf{i} - 7\mathbf{j})$ (d) $\frac{1}{7}(3\mathbf{i} - 2\mathbf{j} + 6\mathbf{k})$
   (e) $\frac{1}{9}(8\mathbf{i} - \mathbf{j} - 4\mathbf{k})$ (f) $\frac{1}{5\sqrt{3}}(7\mathbf{i} - 5\mathbf{j} + \mathbf{k})$

3. $\frac{27}{5}(3\mathbf{i} + 4\mathbf{j})$ 4. $\sqrt{3}$ 5. $2\sqrt{11}$
6. $\sqrt{34}$ 7. $\sqrt{14}$ 8. $\sqrt{195}$
9. $1 \pm \sqrt{39}$ 10. $\frac{3}{\sqrt{14}}(\mathbf{i} - 3\mathbf{j} + 2\mathbf{k})$

11. (a) 3**i** − 2**j** − 8**k**   (b) −**i** + 6**j** + 2**k**
    (c) $\sqrt{41}$ (d) $\sqrt{158}$

12. (a) $\sqrt{242}$ (b) $\sqrt{163}$
    (c) $\frac{1}{\sqrt{507}}(5\mathbf{i} - 19\mathbf{j} + 11\mathbf{k})$

13. 5**i** + 2**j** + 3**k**   14. $\pm\frac{4}{\sqrt{5}}$
15. $\pm\sqrt{2}$ 16. $\pm\sqrt{2}$
17. (a) $11t^2 + 22t + 43$ (b) −1   (c) 32
18. (a) $9t^2 - 10t + 35$ (b) $\frac{5}{9}$ (c) $\frac{290}{9}$
19. (a) $3t^2 - 24t + 86$ (b) 4   (c) 38
20. 73.1°

Exercise 5D

1. (a) -46 (b) -8 (c) 12
   (d) -21 (e) -133

2. (a) $32.3°$ (b) $79.4°$ (c) $112.5°$
   (d) $124.9°$ (e) $95.2°$ (f) $105.7°$
   (g) $138.7°$ (h) $40.4°$ (i) $83.7°$
   (j) $77.0°$

3. 12

4. (a) -3 (b) $\frac{4}{3}$ (c) 10

5. $\pm\sqrt{2}$

6. (a) (i) $41.8°$ (ii) $72.7°$ (iii) $53.4°$
   (b) (i) $27.3°$ (ii) $83.6°$ (iii) $116.4°$
   (c) (i) $68.2°$ (ii) $123.9°$ (iii) $42.0°$
   (d) (i) $107.5°$ (ii) $107.5°$ (iii) $154.8°$
   (e) (i) $29.2°$ (ii) $115.9°$ (iii) $102.6°$

7. (a) $2\mathbf{i}+\mathbf{j}+5\mathbf{k}$ (b) $\mathbf{i}-\mathbf{j}+\mathbf{k}$
   (c) $-2\mathbf{i}+\mathbf{j}-2\mathbf{k}$ (d) $-\mathbf{i}+6\mathbf{j}-9\mathbf{k}$
   (e) $-5\mathbf{i}-6\mathbf{j}+2\mathbf{k}$

8. $-1, 2$

9. (a) $\frac{19}{9}$ (b) $\frac{8}{3}$ (c) $-\dfrac{3}{\sqrt{14}}$
   (d) $\dfrac{11\sqrt{5}}{5}$ (e) $-\dfrac{3}{\sqrt{42}}$

10. $\frac{26}{17}$ 15. $2\mathbf{b}\cdot\mathbf{c}$

16. $\angle AOB = 96.0°$, $\angle BAO = 49.9°$, $\angle ABO = 34.1°$

17. $-\dfrac{14}{\sqrt{255}}$

18. $\cos B\widehat{C}A = \dfrac{13}{\sqrt{806}}$, area $= 12.6$

19. $-\dfrac{2}{\sqrt{17}}$

20. $120°$

Exercise 5E

1. (a) $\mathbf{r} = -\mathbf{i}+\mathbf{j}+\mathbf{k}+\lambda(2\mathbf{i}-3\mathbf{j}+\mathbf{k})$
   (b) $\mathbf{r} = 2\mathbf{i}-7\mathbf{j}-11\mathbf{k}+\lambda(-3\mathbf{i}+9\mathbf{j}+17\mathbf{k})$
   (c) $\mathbf{r} = -3\mathbf{i}+9\mathbf{j}-15\mathbf{k}+\lambda(2\mathbf{i}+\mathbf{j})$
   (d) $\mathbf{r} = 3\mathbf{i}-\mathbf{k}+\lambda(5\mathbf{i}-\mathbf{j}+2\mathbf{k})$
   (e) $\mathbf{r} = 5\mathbf{i}-\mathbf{j}+4\mathbf{k}+\lambda(\mathbf{i}-2\mathbf{j}+\mathbf{k})$

2. $\mathbf{r} = -\mathbf{i}+7\mathbf{j}+4\mathbf{k}+\lambda\mathbf{j}$

3. $\mathbf{r} = 2\mathbf{i}-\mathbf{j}+6\mathbf{k}+\lambda\mathbf{i}$

4. (a) $\mathbf{r} = \mathbf{i}-\mathbf{j}+2\mathbf{k}+\lambda(6\mathbf{i}-3\mathbf{j})$
   (b) $\mathbf{r} = 2\mathbf{i}-\mathbf{j}+7\mathbf{k}+\lambda(5\mathbf{i}+\mathbf{j}-10\mathbf{k})$
   (c) $\mathbf{r} = -3\mathbf{i}-\mathbf{j}+2\mathbf{k}+\lambda(11\mathbf{i}+2\mathbf{j}-5\mathbf{k})$
   (d) $\mathbf{r} = 2\mathbf{i}-6\mathbf{j}+4\mathbf{k}+\lambda(3\mathbf{i}-4\mathbf{j}+7\mathbf{k})$
   (e) $\mathbf{r} = 3\mathbf{i}+\mathbf{j}+4\mathbf{k}+\lambda(4\mathbf{i}-\mathbf{j}-\mathbf{k})$

5. $-3\frac{1}{2}$

6. (a) $\dfrac{x+3}{2} = \dfrac{2-y}{4} = \dfrac{z+5}{1}$
   (b) $\dfrac{1-x}{1} = \dfrac{y+3}{1} = \dfrac{2-z}{2}$
   (c) $\dfrac{-1-x}{6} = \dfrac{y-3}{4} = \dfrac{4-z}{9}$
   (d) $\dfrac{2-x}{2} = \dfrac{y+3}{6} = \dfrac{4-z}{1}$
   (e) $\dfrac{x+3}{2} = \dfrac{4-y}{3} = \dfrac{-1-z}{1}$

7. (a) $\mathbf{r} = \mathbf{i}-\mathbf{j}+\mathbf{k}+\lambda(2\mathbf{i}+4\mathbf{j}+3\mathbf{k})$
   (b) $\mathbf{r} = \mathbf{i}-\mathbf{j}+3\mathbf{k}+\lambda(3\mathbf{i}-4\mathbf{j}-\mathbf{k})$
   (c) $\mathbf{r} = \mathbf{i}-2\mathbf{j}-3\mathbf{k}+\lambda(-2\mathbf{i}+4\mathbf{j}+5\mathbf{k})$
   (d) $\mathbf{r} = \mathbf{i}+4\mathbf{j}-3\mathbf{k}+\lambda(-3\mathbf{i}-2\mathbf{j}-\mathbf{k})$
   (e) $\mathbf{r} = \frac{1}{2}\mathbf{i}+2\mathbf{j}+\frac{1}{3}\mathbf{k}+\lambda(2\mathbf{i}-3\mathbf{j}-\frac{2}{3}\mathbf{k})$

8. $9\mathbf{i}+3\mathbf{j}$ 9. $(4,8,-6)$

10. (a) Don't intersect
    (b) Intersect at $(4,5,9)$
    (c) Don't intersect
    (d) Intersect at $(14,3,-2)$

11. (a) $49.9°$ (b) $68.5°$ (c) $53.8°$
    (d) $47.1°$ (e) $28.7°$

12. Equation of AD is $\mathbf{r} = 6\mathbf{i}+8\mathbf{j}+\lambda(4\mathbf{i}-3\mathbf{j})$
    Equation of BC is $\mathbf{r} = 9\mathbf{i}+12\mathbf{j}+\mu(\mathbf{i}+3\mathbf{j})$
    Intersect at $7\frac{1}{3}\mathbf{i}+7\mathbf{j}$

13. $\mathbf{r} = \mathbf{i}-\mathbf{j}+3\mathbf{k}+\lambda(3\mathbf{j}-\mathbf{k})$; $\mathbf{i}+\mathbf{j}+\frac{7}{3}\mathbf{k}$

14. $4\mathbf{i}+7\mathbf{j}-5\mathbf{k}$ 15. $(1,1,2)$, $70.5°$

16. $\mathbf{r} = \frac{25}{11}\mathbf{i}+\frac{25}{11}\mathbf{j}+\frac{16}{11}\mathbf{k}+\lambda(-\mathbf{i}+3\mathbf{j}+\mathbf{k})$
    $\frac{26}{11}\mathbf{i}+\frac{22}{11}\mathbf{j}+\frac{15}{11}\mathbf{k}$

17. (a) $\dfrac{1}{\sqrt{14}}(\mathbf{i}-3\mathbf{j}-2\mathbf{k})$, $\dfrac{1}{\sqrt{14}}(3\mathbf{i}+\mathbf{j}-2\mathbf{k})$,
    $73.4°$
    (b) $\mathbf{r} = \mathbf{i}+\mathbf{j}+3\mathbf{k}+\lambda(2\mathbf{i}-\mathbf{j}-.2\mathbf{k})$
    (d) $\left(\frac{19}{9}\mathbf{i}+\frac{4}{9}\mathbf{j}+\frac{17}{9}\mathbf{k}\right)$

18. $P(5,-2,3)$, $A(1,0,1)$

19 (a) $(5, 7, 6)$ (b) $26.3°$

20 (a) Equation of AB is
$$\mathbf{r} = 3\mathbf{i} + \mathbf{j} - \mathbf{k} + \lambda(\mathbf{i} + 2\mathbf{j} + 3\mathbf{k})$$
Equation of CD is $\mathbf{r} = 2\mathbf{i} + 5\mathbf{j} + \mu(-\mathbf{i} + \mathbf{j} - \mathbf{k})$
(c) $123°$

Review exercise 2

1 (a) 13 (b) $89°$

2 $\dfrac{1}{3}\left[\dfrac{1}{x} - \dfrac{1}{x+3}\right]$

3 (a) $x\sin x + \cos x + C$

(b) $\dfrac{y}{2} + \frac{1}{4}\sin 2y + C$;

$\sin 2y + 2y = 4\sin x + \cos x + C$

4 (a) $ay + bx = ab\sqrt{2}$

(b) $\sqrt{2}(by - ax) = b^2 - a^2$; πab

5 (a) $\mathbf{r} = 7\mathbf{i} + \mathbf{j} + 7\mathbf{k} + t(\mathbf{i} + 4\mathbf{j} + \mathbf{k})$

(b) $6\mathbf{i} - 3\mathbf{j} + 6\mathbf{k}$

6 $\dfrac{1}{x} - \dfrac{2}{x+1} + \dfrac{1}{x+2}$

7 (a) $1, 0$ (b) $-\sin 2t$ (c) 2

8 $\frac{1}{2}\ln|2y - 1| = \frac{1}{2}\ln|\sin 2x| + C$

$y = \frac{1}{2}\left(1 + \frac{\sqrt{3}}{\sqrt{2}}\right)$

9 $-7\mathbf{i} - 2\mathbf{j} - 5\mathbf{k}$

10 (a) $y = \dfrac{-2}{x^2 + 2C}$

(c) 1 (e) $(-2, -2)$

11 (a) $\frac{1}{4}\sin 2x - \frac{1}{2}x\cos 2x + C$

(b) $\tan y = \frac{1}{4}\sin 2x - \frac{1}{2}x\cos 2x - \frac{1}{4}$

12 $\dfrac{16\sqrt{2}}{3}$

13 (a) $3\mathbf{i} + 4\mathbf{j} + 5\mathbf{k}$, $\mathbf{i} + \mathbf{j} + 4\mathbf{k}$

14 (a) -2 (c) $y + 2 = 0$

(e) $r = -2, s = 4$

15 (b) $a = -\dfrac{3}{2}, \dfrac{dy}{dx} = 0$

16 (a) $\dfrac{2}{x+2} + \dfrac{1}{x-1}$ (b) $\ln 2$

17 (a) $\overrightarrow{AN} = \frac{3}{4}\mathbf{b} - \mathbf{a}$, $\overrightarrow{BM} = \frac{1}{2}\mathbf{a} - \mathbf{b}$

(c) $\frac{1}{2}k\mathbf{a} + (1 - k)\mathbf{b}$

(d) $\left(\frac{4}{5}, \frac{2}{5}\right)$ (e) $4 : 1$

18 (a) $\dfrac{dx}{dt} = kx$

(b) $x = ce^{kt}$ (d) 40.5 hours

19 (a) 13

(b) $\dfrac{1}{105}\left[\dfrac{5}{x+2} + \dfrac{27}{3x-1} - \dfrac{28}{2x+1}\right]$

(c) -6.75

20 (a) $P(-3, 3)$, $Q(3, 0)$

(b) $\left(0, \frac{15}{4}\right)$ (d) 9

21 (a) $\mathbf{s} = \mathbf{r} + \mathbf{t}$ (b) $\mathbf{v} = \frac{1}{2}(\mathbf{s} + \mathbf{t})$

22 (a) $(0, 4)$ (b) $\frac{128}{15}$ (c) $\dfrac{64\pi}{3}$

23 $(y - 3)^2 + (x - 2)^2 = C$
$(y - 3)^2 + (x - 2)^2 = 5$

24 (b) $\dfrac{1}{2}\left[\dfrac{1}{x+1} + \dfrac{1+x}{1+x^2}\right]$

25 (a) $9\mathbf{i} - 2\mathbf{k}$, $3\mathbf{j} - 2\mathbf{k}$; (b) $83.1°$

26 $A = 79, k = 0.023$; 21 min

27 (a) $\dfrac{2}{x+1} - \dfrac{1}{x+2} - \dfrac{1}{x+3}$

(b) $\ln\dfrac{k(x+1)^2}{(x+2)(x+3)}$

28 $e^y = \frac{1}{4}(5 - 2xe^{-2x} - e^{-2x})$

29 (b) $\sqrt{11}$ (c) $35°$ (d) 1.9

30 $\frac{1}{4}\pi^2$

31 (a) $x \geq -p$ (b) $\dfrac{3t}{2}$ (c) 0

(e) $y = x\sqrt{2} - \dfrac{8\sqrt{2}}{9}$

32 $A = 3, B = 2, C = -2$; $\frac{3}{2}\ln\frac{11}{5} + 2\ln 4 - \frac{3}{2}$

33 (a) $\mathbf{r} = (5\mathbf{i} - \mathbf{j} - \mathbf{k}) + t(\mathbf{i} + \mathbf{j} - 2\mathbf{k})$

(d) $3\mathbf{i} - 3\mathbf{j} + 3\mathbf{k}$

34 (b) $\dfrac{4\pi}{3}$

35 $\dfrac{1}{y} = xe^{-x} + C$

36 (a) $\ln(e^x + 6) + C$

(b) $-\frac{1}{6}\cos^6 x + C$

(c) $\frac{1}{2}x^2 - x + \ln|x + 1| + C$

37 (b) 2 (c) $-5\mathbf{i} + 9\mathbf{j} - 4\mathbf{k}$

38 $x\tan x + \ln|\cos x| + C$ **39** $\frac{8}{15}$

41 (a) $1, \frac{1}{5}$ (b) $-3\mathbf{i} + 3\mathbf{j} + 8\mathbf{k}$

(c) $82°$

42 (a) $\dfrac{\pi}{8} - \dfrac{1}{4}$ (b) $\dfrac{a^3}{3}$

43 $(0,0)$, $(-1, -\tfrac{1}{2})$, $(2, \tfrac{2}{5})$

44 $-\tfrac{1}{3}e^{-3y} = xe^x - e^x - \tfrac{1}{3}$

45 $\mathbf{r} = (-3\mathbf{i} + \mathbf{j} - 7\mathbf{k}) + t(4\mathbf{i} + \mathbf{j} + 6\mathbf{k})$
$\overrightarrow{OP} = \mathbf{i} + 2\mathbf{j} - \mathbf{k};\ \sqrt{6}$

46 $g(x) = 2x - 1 - x^2$;
$y = \tfrac{1}{2}\ln[2e^{-x}(1 + x^2) + C]$

47 $-\tfrac{1}{8}$

48 (b) 1.33 (c) 1.58

49 (a) $3s + 4t = 10$ (b) $s = -\tfrac{15}{2}$, $t = -\tfrac{8}{5}$

50 (b) $2e^{\frac{1}{2}} - \tfrac{7}{3}$

51 $(-\tfrac{1}{2}, \tfrac{7}{4})$

52 $y^2 = 2x^2(\ln x + 2)$

53 (a) 5 (b) $3\mathbf{i} + 4\mathbf{j} + 2\mathbf{k}$
(c) $\mathbf{r} = 5\mathbf{i} + \mathbf{j} + 5\mathbf{k} + t(2\mathbf{i} - 3\mathbf{j} + 3\mathbf{k})$

54 (a) $\tfrac{1}{5}\ln\tfrac{8}{3}$ (b) $\tfrac{1}{4}\ln 3$

55 (e) $\dfrac{8c^2}{3}$ **56** $y = 2xe^{\frac{x^2}{2}}$

57 $\tfrac{1}{2}\ln 2 + 1,\ \dfrac{5\pi}{3} - \dfrac{\pi^2}{4}$

59 (b) $-\dfrac{1}{t^2}$ (d) $-\tfrac{1}{8}$

60 $\dfrac{1}{1+x} - \dfrac{1}{1+2x};\ -\tfrac{1}{2}e^{-2y} = C + \ln\dfrac{1+x}{\sqrt{(1+2x)}}$

61 (a) $\mathbf{r} = 5\mathbf{i} + 3\mathbf{j} + t(\mathbf{i} + \mathbf{j} - \mathbf{k})$
(c) $3\mathbf{i} + \mathbf{j} + 2\mathbf{k}$ (e) $4\mathbf{i} + 3\mathbf{j} + 5\mathbf{k}$

62 (a) $\tfrac{61}{192}$ (b) $1 - \dfrac{\pi}{4}$
(c) π (d) $\ln\tfrac{4}{3}$

63 42 **64** -3

65 (a) $0, \dfrac{\pi}{2}$ (b) $12\pi c^2$

66 (a) $\tfrac{2}{3}\sin^3\theta + C$
(b) $e^y - e^{-y} = \tfrac{2}{3}\sin^3 x - \tfrac{1}{12}$

67 (a) $A = 1$, $B = -1$, $C = 2$

68 (a) $\tfrac{2}{3}(x-1)^{\frac{3}{2}} + 4(x-1)^{\frac{1}{2}} + C$
(b) $\dfrac{x}{3}\sin 3x + \dfrac{1}{9}\cos 3x + C;\ \dfrac{\pi}{18} - 9$

69 (a) $\tfrac{2}{9}e^3$ (b) $\ln 2$ (c) $1 - \ln 2$

70 $\dfrac{3}{8(3t+1)} - \dfrac{1}{8(t+3)}$

71 (a) $\dfrac{2r+3}{(r+1)(r+2)}$ (c) $\tfrac{4}{5}$

72 (a) $(2, 2e^{-1})$ (b) $4 - 8e^{-1}$
(c) $\pi(2 - 10e^{-2})$

73 $164°$ **74** (b) $142°$

75 (a) $\dfrac{t^2+1}{t^2-1}$ (c) $\dfrac{29\pi}{12}$

76 $\dfrac{1}{2(1+x)} - \dfrac{1}{2(3+x)};\ y^2 = \dfrac{8(1+x)}{3+x}$

77 $41°$ **78** $2y + 3x = 135;\ 72.9$

79 $\dfrac{4}{3}, \dfrac{16\pi}{15}$

80 (a) $\dfrac{\pi}{32}$ (b) $\dfrac{2\sqrt{2}}{21}$
(c) $\dfrac{2}{3}$ (d) $\dfrac{3 - \sqrt{2}}{5}$

Examination style paper P3

1 ± 2

2 $x^2 + y^2 - 10x - 8y + 16 = 0;\ (0, 4)$

3 $y^{-2} = 2\pi + 4\sin 2x - 8x$

4 $f(x) \equiv 3x^2 - 2x + 2,\ f(x) \geqslant \tfrac{5}{3}$

5 (a) $3\mathbf{i} + 4\mathbf{j} - 12\mathbf{k}$
(b) $OP = 13$ units of length

6 (i)   (a) $-4\cos 2x \sin 2x$
     (b) $\dfrac{-e^{-3x}(\sec^2 x + 3\tan x + 3)}{(1 + \tan x)^2}$
(ii) $y - 1 = \tfrac{8}{3}(x - 3)$

7 (i) $2\tan\dfrac{x}{2} - x + C$ (ii) $\tfrac{4}{3}$
(iii) $\tfrac{2}{27}(3x - 2)(3x + 1)^{\frac{1}{2}};\ \tfrac{100}{27}$

8 (a) $\dfrac{1}{2+x} - \dfrac{2x}{x^2+1}$
(b) $\dfrac{1}{2} - \dfrac{9x}{4} + \dfrac{x^2}{8} + \dfrac{31x^3}{16};\ -1 < x < 1$

List of symbols and notation

The following notation will be used in all Edexcel examinations.

$\in$	is an element of
$\notin$	is not an element of
$\{x_1, x_2, \ldots\}$	the set with elements $x_1, x_2, \ldots$
$\{x : \ldots\}$	the set of all x such that $\ldots$
$n(A)$	the number of elements in set A
$\varnothing$	the empty set
$\mathscr{E}$	the universal set
A'	the complement of the set A
$\mathbb{N}$	the set of natural numbers, $\{1, 2, 3, \ldots\}$
$\mathbb{Z}$	the set of integers, $\{0, \pm 1, \pm 2, \pm 3, \ldots\}$
$\mathbb{Z}^+$	the set of positive integers, $\{1, 2, 3, \ldots\}$
$\mathbb{Z}_n$	the set of integers modulo n, $\{0, 1, 2, \ldots, n-1\}$
$\mathbb{Q}$	the set of rational numbers $\left\{\dfrac{p}{q} : p \in \mathbb{Z}, q \in \mathbb{Z}^+\right\}$
$\mathbb{Q}^+$	the set of positive rational numbers, $\{x \in \mathbb{Q} : x > 0\}$
$\mathbb{Q}_0^+$	the set of positive rational numbers and zero, $\{x \in \mathbb{Q} : x \geqslant 0\}$
$\mathbb{R}$	the set of real numbers
$\mathbb{R}^+$	the set of positive real numbers, $\{x \in \mathbb{R} : x > 0\}$
$\mathbb{R}_0^+$	the set of positive real numbers and zero, $\{x \in \mathbb{R} : x \geqslant 0\}$
$\mathbb{C}$	the set of complex numbers
(x, y)	the ordered pair x, y
$A \times B$	the cartesian product of sets A and B, $A \times B = \{(a, b) : a \in A, b \in B\}$
$\subseteq$	is a subset of
$\subset$	is a proper subset of
$\cup$	union
$\cap$	intersection
$[a, b]$	the closed interval, $\{x \in \mathbb{R} : a \leqslant x \leqslant b\}$
$[a, b)$	the interval $\{x \in \mathbb{R} : a \leqslant x < b\}$
$(a, b]$	the interval $\{x \in \mathbb{R} : a < x \leqslant b\}$
(a, b)	the open interval $\{x \in \mathbb{R} : a < x < b\}$
$y \, R \, x$	y is related to x by the relation R
$y \sim x$	y is equivalent to x, in the context of some equivalence relation
$=$	is equal to
$\neq$	is not equal to
$\equiv$	is identical to *or* is congruent to

$\approx$	is approximately equal to
$\cong$	is isomorphic to
$\propto$	is proportional to
$<$	is less than
$\leqslant$, $\ngtr$	is less than or equal to, is not greater than
$>$	is greater than
$\geqslant$, $\nless$	is greater than or equal to, is not less than
∞	infinity
$p \wedge q$	p and q
$p \vee q$	p or q (or both)
$\sim p$	not p
$p \Rightarrow q$	p implies q (if p then q)
$p \Leftarrow q$	p is implied by q (if q then p)
$p \Leftrightarrow q$	p implies and is implied by q (p is equivalent to q)
$\exists$	there exists
$\forall$	for all
$a + b$	a plus b
$a - b$	a minus b
$a \times b$, ab, $a.b$	a multiplied by b
$a \div b$, $\dfrac{a}{b}$, a/b	a divided by b
$\displaystyle\sum_{i=1}^{n} a_i$	$a_1 + a_2 + \ldots + a_n$
$\displaystyle\prod_{i=1}^{n} a_i$	$a_1 \times a_2 \times \ldots \times a_n$
$\sqrt{a}$	the positive square root of a
$|a|$	the modulus of a
$n!$	n factorial
$\dbinom{n}{r}$	the binomial coefficient $\dfrac{n!}{r!(n-r)!}$ for $n \in \mathbb{Z}^+$ $\dfrac{n(n-1)\ldots(n-r+1)}{r!}$ for $n \in \mathbb{Q}$
$f(x)$	the value of the function f at x
$f : A \to B$	f is a function under which each element of set A has an image in set B
$f : x \mapsto y$	the function f maps the element x to the element y
f^{-1}	the inverse function of the function f
$g \circ f$, gf	the composite function of f and g which is defined by $(g \circ f)(x)$ or $gf(x) = g(f(x))$
$\displaystyle\lim_{x \to a} f(x)$	the limit of $f(x)$ as x tends to a
Δx, δx	an increment of x
$\dfrac{\mathrm{d}y}{\mathrm{d}x}$	the derivative of y with respect to x
$\dfrac{\mathrm{d}^n y}{\mathrm{d}x^n}$	the nth derivative of y with respect to x

$f'(x), f''(x), \ldots f^{(n)}(x)$	the first, second, $\ldots$ nth derivatives of $f(x)$ with respect to x
$\int y \, dx$	the indefinite integral of y with respect to x
$\int_a^b y \, dx$	the definite integral of y with respect to x between the limits $x = a$ and $x = b$
$\dfrac{\partial V}{\partial x}$	the partial derivative of V with respect to x
$\dot{x}, \ddot{x}, \ldots$	the first, second, $\ldots$ derivatives of x with respect to t
e	base of natural logarithms
e^x, exp x	exponential function of x
$\log_a x$	logarithm to the base a of x
$\ln x$, $\log_e x$	natural logarithm of x
$\lg x$, $\log_{10} x$	logarithm to the base 10 of x
sin, cos, tan cosec, sec, cot	the circular functions
arcsin, arccos, arctan arccosec, arcsec, arccot	the inverse circular functions
sinh, cosh, tanh cosech, sech, coth	the hyperbolic functions
arsinh, arcosh, artanh, arcosech, arsech, arcoth	the inverse hyperbolic functions
i	square root of -1
z	a complex number, $z = x + iy$
Re z	the real part of z, Re $z = x$
Im z	the imaginary part of z, Im $z = y$
$\lvert z \rvert$	the modulus of z, $\lvert z \rvert = \sqrt{(x^2 + y^2)}$
arg z	the argument of z, arg $z = \arctan \dfrac{y}{x}$
z^*	the complex conjugate of z, $x - iy$
$\mathbf{M}$	a matrix $\mathbf{M}$
$\mathbf{M}^{-1}$	the inverse of the matrix $\mathbf{M}$
$\mathbf{M}^T$	the transpose of the matrix $\mathbf{M}$
det $\mathbf{M}$, $\lvert \mathbf{M} \rvert$	the determinant of the square matrix $\mathbf{M}$
$\mathbf{a}$	the vector $\mathbf{a}$
$\overrightarrow{AB}$	the vector represented in magnitude and direction by the directed line segment AB
$\hat{\mathbf{a}}$	a unit vector in the direction of $\mathbf{a}$
$\mathbf{i}, \mathbf{j}, \mathbf{k}$	unit vectors in the directions of the cartesian coordinate axes
$\lvert \mathbf{a} \rvert$, a	the magnitude of $\mathbf{a}$
$\lvert \overrightarrow{AB} \rvert$, AB	the magnitude of $\overrightarrow{AB}$
$\mathbf{a} . \mathbf{b}$	the scalar product of $\mathbf{a}$ and $\mathbf{b}$
$\mathbf{a} \times \mathbf{b}$	the vector product of $\mathbf{a}$ and $\mathbf{b}$

$A,\ B,\ C$, etc	events
$A \cup B$	union of the events A and B
$A \cap B$	intersection of the events A and B
$\mathrm{P}(A)$	probability of the event A
A'	complement of the event A
$\mathrm{P}(A|B)$	probability of the event A conditional on the event B
$X,\ Y,\ R$, etc.	random variables
$x,\ y,\ r$, etc.	values of the random variables $X,\ Y,\ R$, etc
$x_1,\ x_2 \ldots$	observations
$f_1, f_2, \ldots$	frequencies with which the observations $x_1,\ x_2, \ldots$ occur
$\mathrm{p}(x)$	probability function $\mathrm{P}(X = x)$ of the discrete random variable X
$p_1, p_2, \ldots$	probabilities of the values $x_1,\ x_2, \ldots$ of the discrete random variable X
$\mathrm{f}(x),\ \mathrm{g}(x), \ldots$	the value of the probability density function of a continuous random variable X
$\mathrm{F}(x),\ \mathrm{G}(x), \ldots$	the value of the (cumulative) distribution function $\mathrm{P}(X \leqslant x)$ of a continuous random variable X
$\mathrm{E}(X)$	expectation of the random variable X
$\mathrm{E}[\mathrm{g}(X)]$	expectation of $\mathrm{g}(X)$
$\mathrm{Var}(X)$	variance of the random variable X
$\mathrm{G}(t)$	probability generating function for a random variable which takes the values $0,\ 1,\ 2, \ldots$
$\mathrm{B}(n, p)$	binomial distribution with parameters n and p
$\mathrm{N}(\mu, \sigma^2)$	normal distribution with mean μ and variance σ^2
μ	population mean
σ^2	population variance
σ	population standard deviation
$\bar{x},\ m$	sample mean
$s^2,\ \hat{\sigma}^2$	unbiased estimate of population variance from a sample, $$s^2 = \frac{1}{n-1}\sum(x_i - \bar{x})^2$$
ϕ	probability density function of the standardised normal variable with distribution $\mathrm{N}(0, 1)$
Φ	corresponding cumulative distribution function
ρ	product-moment correlation coefficient for a population
r	product-moment correlation coefficient for a sample
$\mathrm{Cov}\ (X,\ Y)$	covariance of X and Y

Index